connection drive and desire energy and

ovement pleasure and pain peace and p

action and solace warmth and wonder

nd desire energy and enthusiasm excitement and exercise fu

and pain peace and purpose relaxation and recovery radianc

warmth and wonder balance and beauty challenge an

rgy and enthusiasm excitement and exercise fun and fulfilme

peace and purpose relaxation and recovery radiance an

nd wonder balance and beauty challenge and companionshi

n excitement and exercise fun and fulfilment fear and frustratio

elaxation and recovery radiance and reassurance softness an

nd beauty challenge and companionship completeness an

exercise fun and fulfilment fear and frustration motivation an

recovery radiance and reassurance softness and sensitivit

allenge and companionship completeness and connection driv

ulfilment fear and frustration motivation and movement pleasur

nd reassurance softness and sensitivity satisfaction and solac

hip completeness and connection drive and desire energy an

ration motivation and movement pleasure and pain peace an

ss and sensitivity satisfaction and solace warmth and wonde

connection drive and desire energy and enthusiasm exciteme

ovement pleasure and pain peace and purpose relaxation an

on and solace warmth and wonder balance and beauty energ

WHAT HORSES DO FOR US

WHAT HORSES DO FOR US

Wendy Jago

J.A. ALLEN · LONDON

© Wendy Jago 2009
First published in Great Britain 2009

ISBN 978-0-85131-926-1

J.A. Allen
Clerkenwell House
Clerkenwell Green
London ECIR OHT

J.A. Allen is an imprint of Robert Hale Ltd
www.halebooks.com

British Library Cataloguing in Publication Data
A catalogue record for this book is available from the British Library

The author and publisher are grateful to the following individuals and
organizations for their permission to reproduce material:

Photographs by Leo Jago except for those on pages 93, 137, and 138, by Mick Green,
and page 98 by Wendy Jago
Diagrams on pages 38, 39, 110 and 111 by Rodney Paull, from author's sketches
Edited by Martin Diggle
Designed and typeset by Paul Saunders
Printed in Thailand

For my much appreciated friends, mentors and cheerers-on,
Amanda Prout, Jan Pye and Su Reid, with love.

The special affinity between human and horse: Natasha and Slim.

Contents

Aims and Acknowledgements

This book is about exploration. Writing it has helped me bring together many aspects of my own experience and knowledge – discovering along the way how much more horses do for me than I had realized. I am well aware that I have raised many unanswerable questions; but I hope my observations and ideas may be stimulating and intriguing, or even, at best, clues to greater understanding. What I do know is that in trying to answer my own question: 'What do horses do for us?' I have been drawn into asking some of the deepest questions about my life and its meaning. Horses invited me, and accompanied me, on this exploration, and I have a strong sense that, all over the world and through the centuries, this is a journey they have been making with all of us whose lives they touch so powerfully.

My journey has often been illuminated by the spirit, as much as the technical expertise, of Paul Belasik, Charles de Kunffy, Erik Herbermann and Mark Rashid, each of whom has shone the light of ethics and philosophy upon his experiences as a rider and trainer – and offered me tools for working with my own. I hope that in shining the powerful, yet non-judgemental, spotlight of Neuro-Linguistic Programming (the science of communication which poses, above all, the fundamental question 'HOW?') upon my own experience as a rider, trainer and judge, I may offer a valuable combination of insights and tools to help other riders. Recently, a student said to me that what she and her colleagues really appreciated about my coming to talk to them was that, in sharing some of my fears and struggles, I was 'more like them' than some of the experts they had listened to. She made me appreciate that perhaps this sense of fellow-feeling was as important as the expertise I thought I was there to impart.

Many generous people have shared with me their knowledge, their time, their passion and their friendship, making my explorations companionable as well as thought-provoking. I'd like to thank Aurele Hedley for getting my mind buzzing and for allowing us to attend her equine-assisted therapy workshop; David Harris of Acorns2Oaks for welcoming Leo and me to his Leadership course; Liz Morrison and Andrew Mac-Farlane for letting me join in their LeadChange workshop; and Margrit Coates for making space for us in her healing work and for her support as a fellow-author.

I would also like to thank all the horsy people who responded so frankly to my questionnaire – especially Chris Kent, whose eloquent summation of what horses mean to us graces the endpapers.

At J.A. Allen, Cassandra Campbell, Martin Diggle and Paul Saunders have worked their special magic in encouraging me and producing another gorgeous book. Caroline Burt, who originally commissioned a rather different book, has nonetheless been a valued mentor *in absentia*.

Closer to home, Nikki Green and Gary Keywood continue to be not just friends but supporters, playmates and fellow-explorers, generous with their time, their knowledge and their practical help.

Su Reid, my non-horsy friend for over forty years, entered whole-heartedly into the spirit of the book as I was working on it, giving me continuous encouragement and invaluable advice throughout its development. Now she knows why I have been so obsessed all this time!

Thanks also to Mick Green, who generously devoted a Sunday morning to photographing my husband, Leo: 'Because he's always the one holding the camera, he doesn't get many pictures taken of him riding.' Through Mick's photos, Leo is able to appear in this book as a rider, and we are able to show, rather than just describe, something of his ongoing dialogue with Mouse.

In his other role, as photographer, Leo has bravely undertaken an adventure of his own, largely leaving behind the technical photography he got used to for my earlier books and taking on a much riskier and harder challenge – that of capturing moments of feeling, relationship, communication and connection that help bring the book's most important messages to life.

My special thanks to him, and to you all.

Chapter One

What's It All About?

Who is this book for?

This is a book for anyone who feels a sense of involvement with horses and a commitment to being with them. You may be someone whose life horses have touched quite unexpectedly or accidentally. You may be someone who has regular dealings with them: you might be a rider, someone who works professionally with horses, a teenager or an adult, a therapist whose work is assisted or partnered by horses, a trainer, a breeder, a rider's partner or parent. What you will have in common with the other people on this list is a deep sense of how horses can get under people's skin and how they affect our minds and emotions as well as our bodies. This book is an attempt to show the different opportunities horses offer us and to explain how they can change us for the better.

Start with a person and a horse – there's an affinity. Little girls, old ladies, middle-aged mums, no-nonsense farmers all know it. So do many psychotherapists, social workers, management trainers, people of all ages with physical disabilities, tough thugs, the socially challenged, the sad and the bereaved. Horses attract people from all social classes, all professions, the employed and the unemployed, dwellers in the country and in the town – and they seem to play an important part in emotional healing, social learning and physical rehabilitation as well as simply giving recreational pleasure.

Every kind of person has their own kind of horse. There's a dignity and sense of mutual belonging here which really speaks to me.

What is this book about?

How do horses manage to attract such different people, and to affect them so profoundly? This question has been nagging at me ever since I started *thinking* about my own attraction to them as well as just *experiencing* it. It's not just what horses *do*, but what horses *are*. It's obvious that the powerful attraction horses have for humans rests initially upon things like their size and power; the height and strength they lend us when we are on their backs; the sense that their leaps and their flight are our internal aspirations lent muscle, breath and exhilarating actuality.

But there is even more to it than this. Much has been written about all the technical aspects of being with horses, from breeding, care and training through to advice on the performance of specific disciplines; and recent broader and more questioning explorations have documented the role horses can play in social, mental and emotional as well as physical therapies. Writings on these subjects cited in these pages are listed in the Bibliography at the back of this book.

My interest, and my expertise as someone trained in Neuro-Linguistic Programming (the observation-based system of inquiry that resulted in what has been referred to as 'the users' manual for the brain'), lies in asking and attempting to answer a single, larger question that led to the title of this book and which focuses on the relationship of human and horse in both its most basic and comprehensive senses: *What is it that horses do for us?* This was one of the questions I asked of different people who have horses in their lives, and their replies helped me tease out the answers I was looking for.

Why look more closely?

Isn't being with horses something that should just be enjoyed for itself, without too much analysis? Since you picked up this book, the chances are that you are the kind of person who likes to reflect on what's important to you, as well as experiencing it. Exploring the 'how and why' questions that arise as we go through life satisfies the inquirer in us – the essential scientist in everyone who wants to *understand* as well as simply live their life. I believe that this kind of reflection can add a greater richness to our actual experience, as well as helping to explain it.

When I taught English Literature many years ago, I used to ask my students what reading a poem, a play, or a novel *did for them*. Sometimes my question surprised them, because they came to university with the expectation that studying literature was about analysis: my point was that analysis is really only worthwhile *if it adds something back into experiencing*. Knowing something about grape varieties and where wine comes from helps you to choose wines in future that you're likely to enjoy, and gives you some guidance for experimenting more successfully with ones that are new to you. Thinking about relationships and struggling to understand what makes them go well or badly helps you make them richer, deeper and smoother – and helps you rescue situations faster when they go wrong!

Paying attention to an experience is a powerful way to get even more from it. Not paying attention, on the other hand, means that you get less from it, because, as T.S. Eliot said, doing that is to '*have the experience but miss the meaning*'. Having a conversation with someone when your mind is on something else is a good way to lose the potential value of the conversation – and perhaps alienate the friend as well!

Thinking about what horses and riding mean to you, and how they

The Gift of Opportunity

contribute to your life, can do more than give you a justification for how much time (and money!) you spend on them. It can deepen your pleasure through giving you a more subtle understanding of *just what brings about* that pleasure; and by teasing out the special meaning that horses bring to the unique person that you are, it can help you find your own personal 'recipes' for tuning in to it and relishing it fully. And, as the quality of your attentiveness and the subtlety with which it guides your interaction to your horse both increase, it's likely that your horse will feel more understood and become more at one with you.

What can horses add to your life?

The relationship between human and horse can seem functional and simple, but it can involve so much more. When the climber George Mallory was asked: '*Why do you want to climb Everest?*' he was said to have answered, '*Because it is there.*' When, preparing to write this book, I asked my husband Leo what horses did for him, he answered: '*They challenge my humanity.*'

Leo's reply sounds rather similar to Mallory's, but goes beyond it by implying that being involved with horses not only engages something fundamental in our human selves, but also reveals it more fully to our awareness. Horses, in other words, can offer a mirror to the workings of our deepest selves, giving us the opportunity to know ourselves better and own ourselves more fully.

How do they do this? I think there are two different, yet complementary, ways. First, their reactions reflect how they experience us and what they understand by our actions: what they do when we are with them, when they do it and how they do it, tells us directly and without subterfuge what impression we are making on them. This straightforward mirroring has an unassailable integrity about it, since horses cannot be fooled, bullied or persuaded into deceit.

Second, the discipline we have to impose upon ourselves in order to form relationships with them, gain their co-operation in even the simplest things, ride and train them, acts upon our own selves as well as upon them. As the rider, trainer and writer Paul Belasik has simply and eloquently put it: '*True art always has this integrity, beyond the technical skill. It has the power to transform the self, to enlighten the self, stemming from control of the self.'* (*Riding Towards the Light*, p.97)

The 'mirror' and the 'discipline' achieve between them two contrasting yet related results: they focus us down into the detail of what we feel and how we act, and they open us up to the larger implications for who we are and who we wish to be. As we engage with them we also engage *with ourselves*. Horses can elicit something profound in us without making a song and dance about it (and without us having to make a song and dance about it, either). They can do this because they themselves are authentic and totally without artifice: if we allow ourselves to receive their feedback, they have the capacity to make us more genuine. Where we have issues and weaknesses, they show these up, too, issuing an invitation to self-insight and self-improvement which we can, if we are honest enough and humble enough, accept.

Everyday tasks around horses involve all our senses and require attention – but at the same time make space for us just to 'be'.

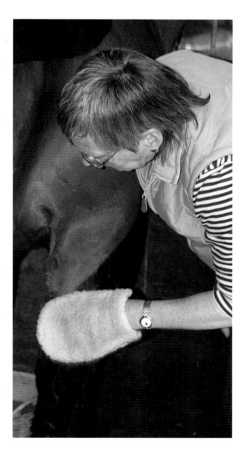

The Gift of Opportunity

Taken together, authenticity, lack of artifice and responses that are not judgemental add up to a very deep level of trustworthiness. When you relate to a being who is like this, you are freed up to engage with the relationship or the task at hand in a very pure way. Just as importantly, as the act of engaging with the horse and the task leads you on to contemplate yourself, the way you will be scrutinizing yourself (and maybe taking yourself on) will be very clean and unencumbered. I believe that, as committed riders, we are much less likely to ignore, block or play down information about ourselves when it comes from a horse than when the same information comes from other people. We can feel private in a horse's presence, and privacy gives us the chance of listening, feeling, reflecting and, if necessary, starting to make changes. When a rider and horse are truly attentive to each other they create a dynamic space of possibility between them, which can be like an experimental laboratory for learning, growth and change.

Individuals and groups

The physical warmth of horses and their frequent gentleness, their readiness to show affection and to live without grudging, show us qualities we admire in others and wish to emulate in ourselves. Engaging with a horse offers not just exercise and the development of skill, but the possibilities of clearer communication with others, less stressful interaction with those around us and, in addition, a greater self-knowledge, deeper personal integration and fuller realization of the even richer self we are capable of becoming.

Making the everyday sublime

Thinking – let alone talking – about the practicalities of horse-care such as mucking out, tacking-up and all the other everyday necessities in such elevated terms as these can make some people feel uncomfortable. I have heard it said that no human action is without the potential to be both ridiculous and sublime. We find it easy enough to categorize some activities as ridiculous; but the notion that everyday acts potentially have sublime significance is one that we can find much harder to take on board. Yet, without such a possibility, how can the influence horses have

There's a quiet companionship and a mutual trust here: it's the spirit of our yard and of many everyday encounters with horses.

on us be as compelling and personally enlarging as, in our heart of hearts, we sense it to be?

Horses compel us to pay attention – to what is going on, to their needs and experience and to our own. We can't *not* attend – and because of this they give us the opportunity of becoming more flexible in our thinking as well as our reactions. We have to keep asking ourselves what to do and why we're doing it and this connects us with beliefs and values that reach back into the past and forward into the future. Through the joys, satisfactions and frustrations of being with them we become strongly aware of what we feel and what we want, but at the same time it's hard to ignore, even in doing the simplest tasks, what *they* feel and what *they* want. Trying to manage the real and potential tensions between these demands on our attention gives us a range of important and exciting learning opportunities.

The Gift of Opportunity

Confronting our selves

Interacting with horses, both on the ground and through riding them, brings us face to face with our selves. Our ambitions, our fears, the kinds of relationships we get into, all tend to surface – as do our assumptions about ourselves and others, our leaps of imagination and acts of generosity. This is the *challenge of humanity* Leo was referring to, which I shall explore to show just how horses can (if we will let them) confront us with our individual qualities, as well as our special bogies, and help us become more effective in working with them.

Alignment and realignment

From my varied experiences over the years as rider, horse-owner, trainer, psychotherapist and self-management coach, I believe that human-horse encounters have the potential for enhancing and developing who we are on very many levels – in fact, upon all the levels of our understanding and being. Managing ourselves around horses means engaging with considerations of environment, behaviour, knowledge and skills, beliefs and values, identity and even our overall sense of purpose and meaning in the world. Between them, these apparently universal, 'hard-wired' mental categories comprise the full range of ways in which we organize both our external experience and our internal understanding of the world. Where we are 'out of synch' with ourselves or others, being with horses seems to offer us the opportunity of aligning or realigning ourselves across these levels, bringing us a new, or renewed, sense of personal harmony.

Personal alignment, sometimes called 'congruence', is one of the key aims of the different forms of therapeutic interaction and training that are offered with horses as guides or partners. When I have talked to the human organizers or practitioners of these therapies, it's been striking how much importance and respect they accord to the horses they work alongside, and how relatively little they claim for themselves. They talk about the horses' perceptiveness, sensitivity, generosity and acumen, instead of their own. In the therapists' experience, they are acting as *enhancers* of the information and guidance offered *by the horses*, rather than taking the lead role as teachers or interpreters in their own right. They rely on the interaction between the horses and the learners/

clients/ patients to offer insights and issue challenges – in much the same way as occurs naturally between horse and human in non- therapeutic situations.

I love the contrast between the dark and the light in this picture. For me, it symbolizes that moment of emergence that starts when someone connects with a free horse in a therapy situation.

Lessons in genuineness

Though horsy people often talk about horses 'being evasive' or 'trying it on', it's a human characteristic to explain the behaviour of others in terms of your own. Horses don't *think* like us because they *aren't* like us. They can certainly exhibit behaviour that seems to us evasive, stubborn or difficult, but this can usually be explained by much simpler and more instinctive motivation than the wish to manipulate others – e.g. by dis-comfort, lack of interest, etc. Being authentic and without artifice goes along with another characteristic: horses by and large act and react in the moment. They may have long memories but they don't bear grudges and they don't make sweeping judgements. They may object to something we do but they don't condemn us as a whole. If we made an ordinary mistake yesterday they don't usually hold it against us today. This is another example we can try to follow in the way we relate both to others and to ourselves.

The Gift of Opportunity

As part of preparing to write this book, I sent a questionnaire to every 'horsy' person I knew, asking them simply what horses did for them in their lives. In reading through the responses, I was struck by how many people described being with horses as 'stress-busting'. Whatever the specific events or feelings that cause it, the feeling of being stressed is essentially a *feeling that things are how they ought **not** to be*. By contrast, the experience of being around horses, or interacting with a specific horse, can induce a state of being that is almost meditative in its intensity and its mind-body benefits. In such a state we can be self-aware – even of our limitations – without condemning or undervaluing ourselves. We can be, at the same time, self-monitoring and self-accepting. Without losing sight of our ultimate hopes and goals, we can be comfortable in the sense that where we have reached and what we have achieved is *enough for now*. This is a state where horses have indeed helped us learn something important about the art of being, and where we experience what it is like to feel fully alive, in a way that, for those special moments, transcends the normal limitations of our existence.

The moral world of the horse

Horses are not perfect, any more than human beings are; yet people's experiences of being around them seem to involve them in issues, choices and decisions which belong in the realm of morality. In my experience, if we are open to their influence, *horses can actually force us to be moral*. As we interact with them, we are continually having to manage the tensions between what *is* and *what ought to be*, whether we are dealing with the pace of training, the demands of competition, the contrast between our ideal and our actual behaviour, or simply comparing what we would like to believe long-term about our horse's goodwill, intelligence and co-operation with the immediate reality of his slowness in learning, 'difficult' behaviour or unresponsive attitude.

Through providing people with challenges like these, horses have helped socially inept teenagers become more aware of other's needs and feelings, enabled people who have been wounded and traumatized to rediscover a value for themselves and a faith in the world, helped those who are physically limited to enjoy and control their bodies more skilfully and effectively than they ever thought possible, and engaged all of us in ongoing conversations with ourselves – and each other – which are philosophical in their nature and can be morally improving in their outcomes. To ask, and to attempt to answer, the question: '*How do they do that?*' is in itself a moral education.

My aim in the book

In this book I want to attempt to answer the question: '*What do horses do for us?*' in a way that is both personal and general. Drawing upon what I know from my study of communication processes, as well as my understanding of classical training and riding, I want to find answers for myself that can also be answers for you, in your own and perhaps quite different circumstances. I want to tease out what is fundamental to the passionate and powerful relationships people have with their horses that helps to explain both the vital satisfactions of everyday caring and the heightened, almost mystical, moments of oneness that can be experienced at different times by each one of us.

I want to explore just how being involved with horses can be so meaningful and potentially such a powerful force for development. It seems to me that something so manifestly intense and compelling as the relationship between human and horse deserves more than anecdote: it deserves *understanding*. As a coach, it's my experience that once we understand *how* someone who is truly skilled does what they do, we are in a position not just to learn from them but to begin to emulate them. Once we understand their recipes in some detail we can begin to make them for ourselves.

So, once we truly understand just *how* horses bring about the powerful and enabling effects on humans that they do, we can – if we so choose – relate to each other in similar ways. We can be powerful yet gentle, honest without unkindness, challenging without aggression. We can enlarge each other's experience and speak to each other's deepest potential just as horses do to ours. Horses so often help make the world better for humans: it's my belief that through understanding how they do it we may be able to make it better for each other.

Chapter Two

The Mirror

Horses don't care about your words: they care about and respond to your actions.

Paul Belasik, *Dressage for the 21st Century*, p.9

'My horse gets so anxious and tense in competition', says the rider attending my 'Make a Difference' clinic. 'He's not too bad in the warm-up, but as soon as we get inside the arena he stops listening to me, spooks at anything even if he's seen it dozens of times before, gets tight in the transitions, won't soften through his back... We can do such nice work at home, but once we get out it's such a different story I'm thinking of giving up.'

This is a story I've heard so many times before. Going somewhere different, seeing lots of other horses, being asked (perhaps abruptly) to do specific things at specific places can make a horse feel excited or worried in his own terms. But it's often hard to tell whether the worry comes initially from him, or whether he's reflecting what he feels from his rider. When we begin to unpick what's going on, we often find that it's the rider's expectations, anxieties and tension that are so powerfully contributing to the equine partner's changed way of going. Riders are quick to blame themselves, but feeling it's all their fault – yet again – doesn't help their confidence. Furthermore, this is not a question of fault or blame – but we do need to accept that all of us bring our own 'baggage' to our riding, and that it can be either beneficial or limiting. Either way, understanding brings us more choices.

In working with many different riders on this kind of problem, I have found that once they identify the personal weighting they are unconsciously adding to the situation it can become possible to take the pressure off themselves – and thereby to change how their horses feel and how they go. Fundamentally, when someone feels that 'performance' has to be good enough to justify the amount of time, energy and money invested in their riding, or feels pressure to do well to please a trainer, to demonstrate competence as a pupil, daughter, partner, adult – human being, even – that person is going to be diverting quite a lot of mental and emotional attention away from their horse. So he is going to feel puzzled, bereft or even abandoned. His tension and anxiety will be mirroring the rider's own.

For so long as such a rider (whether genuinely or with an element of self-deception) buys into the idea that the problems are the horse's, the possibility of making positive changes is severely limited – and is constantly being eroded because nothing fundamental has really changed. Asking and answering the question: 'What does competition – and, especially, doing well in competition – mean *to me*?' can be the key, because it gets to the root of the situation and gives insight and real leverage for improvement. Self-questioning and self-awareness are not the same as self-blame. What we are dealing with is a spiral of actions and reactions; but it is one in which we can intervene and turn around for the better.

Horses reflect back to us how we seem to them, and through their behaviour they show us just what they have understood: they show us not the meaning that we *tried* to communicate but what we actually *did* communicate. They respond to what our bodies tell them through both our deliberate aiding and our accidental signals; and they react to our emotions and often even our thoughts because these, too, are intimated through minute physical changes. Breathing, muscle tone, speed of movement and tiny variations in our normal patterns, all convey meaning to the horse. And if we pay attention to his reactions, if we allow ourselves to take in honestly and humbly the information he gives back to us, we can learn a great deal. We can learn to understand and manage him better – and we can also learn more about ourselves. Understanding how horses mirror us, and how we can make sense of what we see in that mirror, is what this chapter is about.

The Gift of Reflection

In her amazing book *Animals in Translation*, the autistic psychologist and animal behaviourist Temple Grandin says:

> I do know people can learn to 'talk' to animals, and to hear what animals have to say, better than they do now. I also know that a lot of times people who can talk to animals are happier than people who can't. People were animals, too, once, and when we turned into human beings we gave something up. Being close to animals brings some of it back.

> (Temple Grandin and Catherine Johnson, *Animals in Translation*, p.307)

In her impeccable research background, as well as her wide practical experience in helping people manage animals more effectively and more humanely, Temple Grandin is a long way from woolly sentimentality. The very 'handicap' of her autism gives her a special take on how people and animals perceive the world and how they try to communicate their understanding.

> I'm different from every other professional who works with animals. Autistic people can think the way animals can. Of course, we also think the way people think – we aren't *that* different from normal humans. Autism is a kind of way station on the road from animals to humans, which puts autistic people like me in a perfect position to translate 'animal talk' into English. I can tell people why their animals are doing the things they do.

> (Ibid., pp.6–7)

In explaining to us how animals see the world, Temple Grandin shows us how they get the information about us that they do. She shows us, in other words, that what we perceive in the mirror they offer us is valid. In this chapter I want to look at three aspects of the mirroring process.

- First, how a horse's behaviour can tell us about our own (the process of giving and receiving feedback).

- Second, how we can blur and distort that feedback's meaning (deliberately or inadvertently) and how we can ensure that we receive it more clearly.

Is Shane mirroring her horse's quiet concentration, or vice versa? However it began, what we see here is a 'virtuous circle' of mutual attentiveness.

- Third, how valuable it can be to understand and work with feedback, not as a single action but as a sequence of actions and reactions between beings who, by virtue of their connectedness, are part of the same *system*.

The feedback horses give us

Horses are great observers. They have acute eyesight and sensitive hearing, which enable them to monitor what's around them. They are, of course, historically prey animals, so they filter information differently from us and are predisposed to swift flight responses. When we are riding them, however, it's their physical or kinesthetic sensitivity that governs their responses and provides the essential basis for the way they mirror us. It has been observed that the point of closest contact between horse and rider is where their spines – *and therefore their major nerve and information systems* – come together. This gives the potential for the really rapid transfer of information between one partner and the other, and I think it may be one significant means whereby, at times (especially in established partnerships), the horse can seem to respond to the rider's thought or intention rather than a deliberate action. It's almost as though the thought-message jumps across that gap from flesh to flesh, rather as our own instinctive reactions to stimuli such as sharpness or extreme temperature involve a jump across from receiving nerves to reacting impulses without first passing through conscious awareness. I don't know whether this has been studied scientifically, or indeed if it could be, but it's certainly been well attested to by the experience of many riders. As riders, this challenges us to control not merely our bodies – difficult enough – but also our very thoughts…

I am judging an affiliated Medium level dressage competition. It's early January and freezing, so by the time my class starts in late afternoon most of the people entered have given up and gone home: only three are left. As it happens, they are very different: the contrasts between them make me think ahead to the chapter on horse and rider mirroring that I am about to write.

The first horse is attractive and well set-up, though his legs seem long for his body. He is rarely 'through' in the neck. He is enthusiastic and goes into canter every time his rider asks him to go forward. Each time he bounds off she calls him back. They progress in a series of stutters, rocking forward then back again. There's a kind of ricochet effect going on between them. Their marks oscillate between 4s and 6s.

The second horse looks rather solid and very genuine. His gaits are not extravagant, but he is working willingly and in balance, and he seems to enjoy listening to his rider and doing his best for her. They are both 'soft', and their work is fluid. The rider's elbows are close to her sides, so her hands are kind, and her lower back follows her horse's movements closely. She is able to help him show all three walks (collected, medium and extended) clearly – there's a real difference in the height of his steps and the reach from his shoulder, and his back really swings and participates appropriately in each version of the gait. They have a lot of 7s and a couple of well-deserved 8s.

The third horse clearly has expressive, elevated natural gaits, and seems to have been well educated. He looks potentially talented, but his

The Gift of Reflection

tail isn't swinging freely in that S-shape trainers and judges hope for, and though his balance in the small circles and lateral movements is good (he gets 8s for some of these) he tends to brace his neck and isn't swinging freely through his back. His rider's arms aren't 'attached' to her body – her elbows are pretty straight – so her hands both cause the movements of her horse's head and reflect them: the contact is mutually reactive. As for their backs, I can't tell whether the horse's 'holding' is causing the rider's stiffness, or the other way about. They are not easy and flowing. Despite the accuracy and technical proficiency which earn them steadily reasonable marks, the middle horse beats them by a whisker.

These horses are all giving their riders feedback. If each rider were willing to take her horse's way of going as a commentary on her way of riding, by working on herself she could probably help him perform better and feel more comfortable in his mind and body.

The first rider could strive to make her aiding subtler and 'quieter' so that her horse could keep something of his expressiveness and his forward flow even when he bounds off or is making downward transitions between and within gaits.

The second rider might be encouraged by the harmony she has with her horse and begin to ask for more expression and more elevation (both something of a risk because they might disturb that safe and pleasant evenness they have at the moment).

The third rider could ask herself what she is doing to disturb the contact, and seek to stabilize and elasticize her hands and arms. She might also remind herself that, since her horse is naturally talented and they are both well-educated, they could afford to relax more: it's likely that even if they lost a few marks through being less precise they would gain them all – and more – back for being softer and more flowing.

Having made these observations, it's important to allow for the fact that your horse sometimes behaves not like a mirror but just simply like a horse! Sometimes that stiffness, that stubbornness, that lack of attention are indeed coming from him. Wishing he could still be chasing his friends in the field, feeling tired after a long hack, feeling bored with schoolwork are all possible reasons for his mind and his energies to be less available than you'd hoped. Every ride is a new beginning – and not all of them are ideal!

Nonetheless, it always pays to ask yourself whether your horse's behaviour now, today, or in general is offering you feedback on your own behaviour. This requires good observation, good 'feel' and considerable humility. But the willingness to look at yourself honestly through the mirror of your horse opens up a world of possibilities on a daily basis. Every time you ride you have a potential teacher with you – your equine friend. In paying attention to him, you become sharper-witted, more observant and more versatile in problem-solving. Even if you have a human trainer, you become less dependent on them and more able to continue and refine your own education.

How we distort feedback

Feedback is only useful if it is received fully and openly. But it still has to be made sense of, and that inevitably means that it has to be simplified and filtered. Once you filter, however, you become liable to error. It's something of a Catch-22 situation. Neuro-Linguistic Programming has helped us identify the commonest ways in which people filter available information – a process that means that what they *expect* to see or hear actually shapes how they interpret the information being 'sent' to them. There are three main filtering mechanisms:

- Deletion
- Distortion
- Generalization

Each has a positive value in life because it helps us simplify the mass of information available to us – but, by the same token, if we forget that we are always tending to simplify, we can end up not noticing things that are important – or even vital. We can end up reaching conclusions that are far from the truth.

Riders are no different from anyone else in the way they filter the information coming from their horses. Mostly, they do it unintentionally – after all, who would really seek to block information that might help them ride more effectively and feel more at one with their equine partner? But they do it, all the same! Let's explore what prevents us from receiving feedback 'cleanly', and what we can do about it.

Deletion

Deletion means that you ignore some of the information that is available. There may be all kinds of reasons, both conscious and unconscious, for doing this. Deleting can also be about choosing to emphasize some things at the expense of others: if you focus in on one thing it is often at the expense of something else you could or should be noticing. Perfectionists, for example, may effectively delete things their horse is doing well because they are focusing on how he could have done better, didn't co-operate fully; on how they lost marks, or how things didn't go as well as they'd hoped today. It's as though positive information isn't as valid – maybe not valid at all. So, effectively, they haven't actually received it.

Another clinic. The rider introduces her horse in a way that is affectionate yet at the same time sounds apologetic about him. 'He's such a good boy, yet lots of people would write him off because he's only a cob.' She would like him to be steadier on the bit, and is inclined to blame herself for having 'restless hands'. Yet I sense that she is producing a 'problem' she thinks I can help with rather than one that's actually deeper, more personal and more troubling to her. While we are playing together I learn how much she has already achieved with her 'broken-down, written-off, uneducated' horse. And I also sense how little credit she gives herself for how she's helped him settle, blossom and regain his confidence. In focusing on what still remains to be done she seems to have lost touch with the joy and pride she could justifiably be feeling. I think it's this lack of joy and loss of self-confidence that she is really wanting help with. I decide to bypass the 'official' problem for a while and remind her through some 'new' learning how good a partnership they already have and how well they communicate with each other. She hasn't realized how much her horse already listens to her seat – but when I ask her to ride serpentines 'on the buckle' and make downward transitions just by sinking her weight more heavily down through softened ankles she soon finds out how well the two of them understand each other. 'I never knew we could do that' she says. Both their faces express delight.

Of course she was right in thinking they need to improve their rein contact – but in focusing on what was missing there she was largely deleting the positive elements of her riding and distorting the true picture of the partnership. And in doing so, she made herself feel less competent, and much less joyful, than she had needed to; and so she and her horse were having less fun together than they could have.

Distortion

Distortion is what occurs when the balance of the 'truth' has been skewed because of the perceiver's assumptions, beliefs or wishes. Distortion can sometimes be helpful as, for example, in the case of a rider who recently told me how pleased and proud she had been to get a mark of 59% 'because we were in such good company'. As NLP also shows us, the way you frame something mentally is everything!

On the other hand, interpreting riding difficulties as being the result of your horse's inadequacies (physical or mental) often also distorts

However well-trained they are and however effective we are as riders, we have to remember that horses are always horses! Leo caught the moment when Ollie was surprised by something in the distance, lowered his quarters and made a graceful half-pirouette in the opposite direction…

because it leaves out a number of important facts. It distorts by emphasizing his contribution (rather than, or more than, your own), and doesn't take account of the fact that horses think differently from humans and so have very different motivations. As Temple Grandin says in *Animals in Translation* (pp. 51 and 57): 'Normal human beings are blind to anything they're not paying attention to: they stop seeing the details that make up the big picture, and see only the big picture instead … Animals see all the tiny details that go *into* the big picture.' On the day in question, your horse's attention may well have been elsewhere – perhaps on things you had overlooked or hadn't even noticed.

Generalization

Generalization, the third simplifying mechanism, filters information by taking one example and assuming it applies to others that are in some way like it. Mares are tetchy… horses are fresh when they are changing their coats… Such examples are not so different from the assumptions people make about other races, children, women, businessmen… Once you allow yourself to look at things like this, with blinkers on, you are going to miss much of the rawness of reality, with its significant contradictions, inconsistencies and unexplained occurrences.

Another rather specialized kind of generalization is when you look at the information in front of you and judge it according to how it '*should*' be, '*ought to*' be or '*must*' be. You are attempting to bring living experience within the constraints of a particular belief or set of beliefs. You can test for generalization by simply asking questions like: 'Are *all* mares tetchy?' 'Are horses *always* fresh when they are changing their coats?' 'Should you *never* allow a horse to do [whatever it is]?' If the answer to any of these is 'No', you are guilty of generalizing!

Generalizations are not always misleading and unhelpful, of course – but it is worthwhile knowing whether they are skewing how you are making sense of the evidence available to you.

I'm thinking about our horse, Mouse, an ex-eventer who was very obedient but very stiff when we got him. Tall and relatively long-backed, he moved like a plank with twinkling legs: no bending, no undulation, no connection through his top-line. He had been taught to assume a 'correct' shape, but was lacking the muscle or the training that would

The Gift of Reflection

have enabled him to carry himself or work with athletic freedom. Judging him before we ever knew he was for sale, I wrote: 'Lovely attitude – just needs to soften and connect through his back more'.

After a year's unhurried work, Mouse knows that he has a back, and sometimes what it's for. But he and we have reached a kind of plateau. Sometimes he works 'through' when I ride him, and sometimes not. I realize that I'm engaging in *distortion*: I'm thinking: '*He finds it easy to bend to the right and hard to bend to the left.*' His behaviour tells me this – but actually, if I'm open to its full mirroring, I can take the feedback more personally – and more usefully. *I* find it in some ways easier to bend to the right (as witness, I tend to collapse my right hip) and harder to bend to the left. Once I realize I'm distorting the information Mouse is giving me, I can pay better attention and do more with it. I observe that when we're on the right rein we bend more easily – but we also tend to cut in. I ask myself the uncomfortable question: '*How am I different when **I'm** on the right rein from when I'm on the left rein?*' Then I realize that I'm sliding out to the left, so my inside seat-bone and inside leg can't be effective, so of course I can't support Mouse and help him balance. Result, *we* fall in (not just *him*).

In this example of personal schooling, I realize I'm also *generalizing*. I'm telling myself that, because Mouse is less consistently on the bit in trot than in walk and canter, we should spend time sorting that out: the implied generalization is: *you should focus your schooling on problem areas*.

But I've forgotten that the meaning I attach to being 'on the bit' represents 'submission' in human terms – but something very different in a horse's terms. Of course I understand from my classical training that acceptance of the bit is not an end in itself but the result of something else – a biomechanical process that happens naturally when the horse is connected through the whole of his top-line. As he tucks his pelvis, supports his weight from behind, elevates his spine by contracting his abdominal muscles, he will be engaging a whole system of connected postural muscles that allows his head and neck to relax and be carried from above rather than propped up from below. Going onto the bit is a natural by-product of engaging this postural process. In terms of the horse's instinctive understanding, being required to go on the bit by any means other than the engagement of his postural muscle system is an act of manipulation, rather than assistance.

I've forgotten one of the most useful generalizations (some can be helpful!). It's Einstein's comment that a problem can't be solved in its own terms: in order to solve a problem you have to change the way of thinking in which it's embedded, or the context in which you under- stand it. So – *who says we have to solve trot problems by more trotting?* If the problem is about lack of looseness and 'throughness', why not start with where these qualities come easily? I remind myself that part of the problem is likely to be me. As a judge, I often notice that horses are tighter and more inclined to come off the bit in the trot than in the canter. Most riders find it easier to remain seated and swing in the canter and to keep their hands steadier in a following contact: why should I be any different? It also seems easier for a horse's back muscles to work more comprehensively in canter than trot. What might Mouse's variable 'throughness' and varying acceptance of the bit in trot be telling me? Something like: *help him get the habit of 'throughness' and consistent contact in the canter – and work on your own inconsistencies in the trot!*

In the schooling sessions that follow, we canter a lot. Our balance and our contact improve. We increase engagement through pliés (leg-yields from the short-side centre line towards B and E), through spiralling in and out and through shoulder-fore. When we are loosened up, we try to maintain our softness and 'throughness' in trot. Mouse begins to ask for a deep, stretched contact in all three gaits. We are happier. We may have cracked it – not necessarily the 'throughness', not the acceptance of the bridle, but the filter of the 'ought' that was cramping our progress.

The horse and rider as a system

'What is a system? A system is something that maintains its existence and functions as a whole through the interaction of its parts.' (Joseph O'Connor & Ian McDermott, *The Art of Systems Thinking*, p.xiii)

Riders and their horses affect each other, whether they are long-established partnerships or temporary ones (as in a one-off hack or lesson). They affect each other physically, as their bodies impact on and make demands of each other; they affect each other mentally, as they

Soft and through… But the fragility of the connection we have achieved in this session can be seen in my very careful, almost passive, body language and in Mouse's thoughtful expression. Let's not risk disturbing it, eh?

continuously test each other's understanding of what they are trying to communicate; and they affect each other emotionally, as they pick up and often amplify each other's emotional states.

Therefore, as a riding partnership, you and your horse together make up a system – yet you are each also systems in your own right. To quote O'Connor and McDermott from the same page as above: 'Your body… consists of many different parts and organs, each acting separately yet all working together and each affecting the others.'

This doubling up makes riding one of the most complex systemic activities; and of course it's one in which the rider carries the responsibility for monitoring, reinforcing and altering how that complex system is functioning. Let's look at how this can work in practice.

I go to teach my friend Roy on his Advanced Medium horse, Cori. He has a regular trainer, Marion, whom I know well, and she and I seem to have reached an unspoken understanding that she helps Roy to work on Cori while I work on Roy. Each of us trusts the other to do what she is best at – it's a good arrangement and Roy appreciates and learns from us both.

Today, Roy tells me that in his most recent lesson with Marion they were working towards getting Cori to become lighter in front and less inclined to overbend and lean on the hand. As Roy tells me about this, he is riding around me in the school, and something draws my attention to his lower leg and foot position. His heels are up and, when he is not actively aiding Cori, his lower legs tend to wave about loosely. We have remarked on this before – Cori at 15.3 hands is quite small for Roy's long legs, and I think this is why Roy brings his heels up to reach him. Of course this weakens the effectiveness of Roy's seat. I ask him to lower his heels, and remind him that he can best aid Cori with the upper part of his calves, since these are the only parts of his lower legs to be naturally against Cori's sides. Then I remember once riding and trying to 'get through' to a client's horse who was so tall and massive that my lower legs hardly touched him: in the end the only way I could steer him was with my thighs and knees! This might work for Roy too – though for a quite different reason! I suggest that he tries using his thighs and knees to shape and steer, so that his lower legs and heels can stop trying to do all the work and just remain stretched down under him. In order that he can use his thighs and knees to best effect we need to reposition Roy, making sure that he turns his whole legs inwards from the hip joints so that his thighs can lie flat against the saddle – a position that is not only new but really quite difficult to maintain: the human leg seems much happier turning outwards when it's on a horse!

The results seem miraculous! Not only do Roy's legs remain correctly positioned for virtually the whole of our session but, as a result, his seat becomes deeper and more stable. He doesn't bounce in the sitting trot or canter, even though he says he is feeling stiff today. More amazing still, Cori hardly overbends at all, even in downward transitions. Roy plays with spirals, with canter-trot-canter, with simple changes and then flying changes. Everything is crisper, lighter, more balanced. Cori is softer in the neck and much more 'uphill' – and Roy proves that this is genuine self-carriage by giving away the inside rein from time to time. Cori isn't leaning any more, nor is Roy holding or 'hoiking' him up.

The Gift of Reflection

This is what can happen when we take a horse's behaviour as a mirror, seeking to learn from it and to use that learning systemically. Roy accepted that his leg position would be having an effect on his seat and his balance – and thus giving Cori a message he actually didn't want him to have. As Roy changed, Cori reflected the differences. As Roy perceived that changed reflection (greater balance, increased lightness, crisper transitions) he was able to give even lighter aids. Because Cori was in good balance he was able to respond quicker and better. The mutual mirroring became one of harmony, balance, and increasing precision.

We could explain this by saying that Roy had changed the pattern of the systemic interaction between himself and Cori. For as long as any of us bought into the assumption that Cori 'likes to overbend' and 'tends to go onto the forehand', the major thrust of our interventions would be directed to 'stopping him overbending' and 'getting him to engage more and lighten in front'. Once we changed our thinking and asked what kind of commentary Cori's behaviour was giving on Roy – how he was mirroring his rider, in fact – then we discovered the ways in which Roy was inadvertently *telling* Cori to overbend and disengage behind. Seen systemically, the problem wasn't one of what Cori *liked*, or *wanted*, or *preferred* to do, but one of what he *understood from his rider's way of going*. Changing Roy's way of going enabled Cori to change *his* way of going, in the most delightful, subtle and relatively effortless way. I say *relatively* effortless – in addition to the leap of faith required to have a go at a different way of aiding, Roy also had to make the physical effort of repositioning himself and sustaining that repositioning. The proof that he had done this was that the marks made by his saddle on his white breeches were in a quite different place from usual – not underneath but more towards the top and front of his thighs and knees.

It is probably because we connect with them systemically that horses can influence us so much. And of course this happens whether or not we are analysing the process consciously. But once you do become aware that you are part of a system, you have so many more choices, both in the way that you think and the way that you behave. Much of this book rests on the belief that awareness opens the way to more, and better informed choices:

We can change the way we think for today, looking first at our part in the system, then at our mental models, taking time delays into account and realizing that we do not really escape the consequences of our actions. When we change the way we think, we will change the way we behave in a reinforcing loop, which in turn will change the way we think… And this may lead us to wiser counsels.

Joseph O'Connor & Ian McDermott, *The Art of Systems Thinking*, p.229

The Gift of Reflection

Chapter Three

The Discipline

The player of the Inner Game recognizes increasingly that *all* moments are important ones and worth paying attention to, for each moment can increase his understanding of himself and life.

Timothy Gallwey, *The Inner Game of Tennis*, p.125

Re-approach everything you know with wonder and curiosity. Study yourself, humans in general, and then your work organization. Learn from people through observation and interaction. Learn to see, listen intently with your eyes and ears. Allow your mind to accept and receive messages you never realized existed.

Monty Roberts, *Join-Up: Horse Sense for People*, p.238

Riding, being with and caring for horses all invite us to make a serious commitment both to them and to ourselves. Such a commitment can require much more than the time we spend with them, and can involve a discipline as much of the mind and spirit as of the body. Because of their power and their rapidity of reaction, horses more or less compel us to pay attention and be present in here-and-now awareness. However, since our minds are also able, in a moment, to flip back into the past, forward into the future or sideways into thoughts related or unrelated to our immediate surroundings, we can always be led astray.

To put this another way, horses make us aware that, all the time, we have the choice between being present here-and-now or mentally present somewhere else. We have the same choice in every other moment of our lives, of course, but we are not usually reminded of it so directly, clearly and without moral overtone, as when we are in the presence of horses. Absence of mind while driving may cause an accident – or just give you a nasty moment or two when you realize what might have been… Losing attention when someone is talking to you may make the other person angry or cause them to feel belittled or discounted; and because your mind is elsewhere you may not even realize the effect you're having. But, even in these situations, more is actually at stake than control or someone's good opinion of you. For however long your mind is detached from what you're experiencing you're losing something from your own possible experience of being alive. As Neil Postman and Charles Weingartner put it:

> It would appear that… People 'happen' as wholes in process. Their 'minding' processes are simultaneous functions, not discrete compart-ments. You have never met anyone who was thinking, who was not at the same time also emoting, spiritualizing, and for that matter, livering.

> Neil Postman & Charles Weingartner, *Teaching as a Subversive Activity*, p.87

This reminds us that we are experiencing life in many ways at any one time. For most of us, some kinds of experience are harder to tune into than others: 'livering' might be one example – though most people sense pretty accurately when they are 'off-colour', which may be one way we are reminded of how well our 'livering' is going! Some kinds of self-monitoring seem to come naturally – sometimes so much so that we don't even realize how remarkable they are. For example, some years ago I noticed that if I guessed my weight before stepping onto the scales the figure that came into my mind was virtually always the right one – to such an extent that if I was hesitating between two amounts only a pound apart the scales would flicker between those two amounts before settling to one of them. My mother rarely wore a watch because she always knew what time it was. You can probably think of examples from your own experience.

How can we make the most of all the 'minding processes' that are available to us? If we wish to, and with practice, we can develop greater sensitivity to the different strands of our functioning, and notice even

The Gift of Presence

small fluctuations. I'm not arguing, of course, that we should seek to have *all* of our awareness in the present: to do that would be to forfeit some of the varied richness and intensity that can be ours, and to compromise what it means to be human in a different way. Rather, I'm talking about developing a kind of multi-stranded awareness that consists of an ongoing *relationship* or continuing dialogue between inner and outer experience, between present, past and future, which essentially means monitoring and managing many possible calls on your attention. We could think of this as a demanding act of juggling, or compare it to following themes and instruments through a complex piece of music. Neither of these, though, makes such a range of demands on our ability to multi-task mentally, emotionally and physically as the demands made of us by a horse!

This, no doubt, is what underlies the helping and healing potential of being around and working with horses: the demands they make on us are what create possibilities to become more aware, and more self-aware. By calling us into the present, horses help us create in ourselves a discipline of attention and reflection. In this chapter I'm going to explore just how this happens, outlining some useful skills that allow us to become more deliberate and self-governing in our use of awareness. These are:

1. Stepping into others' shoes.

2. Applying what you learn to other situations.

3. Using the feedback you get from the immediate situation to ask even bigger questions.

Stepping into others' shoes

There are a number of standpoints from which we can perceive any situation and any participant within it – including ourselves. Learning to shift the standpoint from which you experience things – rather like employing computer graphics to rotate an object or take you on a virtual tour onscreen – opens up a greater understanding of your own and others' experience in any given situation (individual 'realities'), and also connects you with an objective viewpoint that helps you benchmark 'reality' in a more complex and comprehensive way.

People who love their pets usually feel like they have a pretty good idea what an animal needs to have a good life. The basic necessities for life for pets are the same as they are for us: food, safety, companionship.

That's a good start, but if that's all you know about animals you can still get into trouble… Anyone who's gone out and bought himself a Border Collie… is missing one big item from the Border Collie list, and that is a job.

Temple Grandin and Catherine Johnson, *Animals in Translation*, p.179

When we are around animals, it's easy to assume that essentially they're rather like us. But sooner or later, like the Border Collie, they will react in a way that confronts us with the fact that they are not like us: *they are like themselves*. The Border Collie gets irritable, difficult or even dangerous because *he has nothing purposeful to do*. The horse threatens to kick out when you try to pet him at feeding time *because he fears you may steal his food*. He retreats to the back of his box away from your small child with her quick, unpredictable movements and high-pitched voice *because he has learnt that high-pitched sounds and rapid movements signal danger*. Out hacking, he refuses to go past the blue hosepipe by the barn that wasn't there yesterday: *it might be a snake*. He is not being difficult or nasty: he is just being a *horse*.

When he reacts like this, we have no choice but to respond. One option, of course, is to get cross with him, but we are likely to find that this simply makes him more frightened and less inclined to trust us and do what we want. Another option is to try to imagine ourselves into his shoes ('What must he be thinking and feeling to respond like this?') and to let this govern how we react. When we attempt to put aside our own reasoning and adopt his, we are making an imaginative transfer from our position in the world into his. We are setting aside our own, *first position* or viewpoint and trying on a *second position*, which in this case is his.

Martin Buber developed his I-Thou theory because of a horse he came to love. The I-Thou theory helps us grasp the different styles humans use to relate in the world. When individuals develop a capacity to genuinely

The Gift of Presence

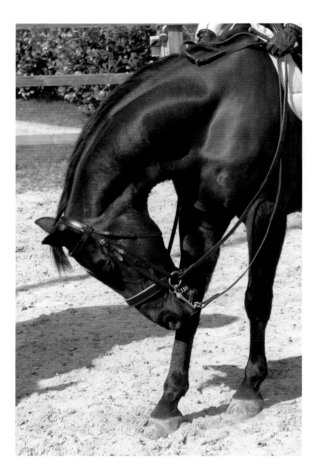

Ollie has developed a habit of stopping and rubbing his nose on his knee. Is he dealing with a physical itch, or buying himself a moment off from the discipline of concentrating? We don't know yet, so Nikki allows him his moment before asking him to get back to work.

reach out and take an interest in another, their style of relating is healthier and more interactive. They find that listening is the key, and they accept others' worlds and feelings without comment.

A.V.R. and M.D. McCormick, *Horse Sense and the Human Heart*, p.102

Most of us are used to inhabiting that first, self, position and feeling that we have a right, more or less, to all it means to us; but when a person's history has given them pain, misery and trauma, or when the significant people in their lives have denied their right to count as individuals, being suddenly confronted with the reality of a horse doing what is, for him, first position, can be a tremendous, if challenging, opportunity – because he has to be responded to not tomorrow, not next week, but right now!

The Gift of Presence

In responding, however, the sense of self that someone has established through their individual life history can often stop them communicating clearly. Horses understand energy, and when we are fearful, doubtful or lack confidence, what we intend and want doesn't come through so well. Horses respond to the kind of energy these feelings give off by distanc-ing themselves, emotionally and sometimes physically. What they do respond to is *authenticity*. As Wyatt Webb puts it in his book *It's not about the Horse* (p.101): 'A horse is consistent in his awareness – pure as he can be, totally sophisticated and always in the moment. A horse knows what to do every single time. It just depends on how clear *you* are as a human being.'

The central mechanism through which horses mirror people and can offer them the opportunity to make therapeutic changes, is that they are so firmly placed in their own first position that through their immediate, unequivocal responses they are able to expose the ways in which we have learnt to be uncomfortable in ours:

> When we work with horses, they help reflect the areas in our lives that block us or bring us into conflict with ourselves and others. So, instead of making it about the horse (or somebody else) we can take the oppor-tunity to change…
>
> Once we've seen what our problems are, we can ask ourselves how we normally deal with situations like the one the horse brought up. Now, if we've tried what we knew and it worked, great. But if it didn't work, the opportunity for change has arrived: *let's not do that anymore.*
>
> Wyatt Webb with Cindy Pearlman, *It's not about the Horse*, p.140

Once we become more aware of what belongs to us in our first position, we are more able to take up the invitation to understand what is going on for the other: to step into second position.

And there's yet another position from which we can view things – the position of the neutral yet attentive observer. This is *third position*. When we actually *are* the observer, this is relatively simple. We are watching an ongoing situation and trying to make sense of it. What's going on, maybe also *why* is it happening like that? We may be making a kind of running commentary to ourselves in the privacy of our own head, or even having a discussion with ourselves. More difficult as a discipline,

The Gift of Presence

yet even more rewarding, is cultivating the ability to run a third-position observer-view of an ongoing situation *alongside our first and even second position views*. It goes something like this: here I am being myself (first); I'm also trying to be as attentive as I can to the signals I'm receiving from the horse (or the person) I'm interacting with, asking myself, 'What's it like for them?' (second) ; and, in addition, I'm asking myself questions like, 'Am I doing the right thing here?' 'What opportunity/challenge is this situation offering me/them?' 'Did that work?' (third). Here's a recent example.

I offer to help my friend Nikki school her horse, Ollie. We often do this for each other – and we both use the word 'help' rather than 'teach'. On this occasion Ollie isn't exactly energetic – he's not unwilling but his walk is a bit of a slouch and his trot lacks energy. As I notice this (first) I find myself wondering what Ollie might be feeling (second). He's used to the school and to starting off with a bit of walking and some relatively collected trot – might he be a bit switched off by the familiarity of it? He looks bored. I ask myself what would be more fun for him? What about some forward, expansive, trot and canter? More like a hack (he loves hacking). I suggest this to Nikki. This is much better – much more energy and flow. I am pleased with myself (third position) – as a coach I seem to have hit on a good strategy.

Later on in the session, Nikki starts asking Ollie for some half-pass in trot. It's correct but without much expression, and he's not covering as much ground sideways as I think (first) he could. How could we help him become more athletic without badgering him? Could I find a way to put him in the position of doing what we want (third) – but as if it was his own idea? (second). I remember Kimberley, my first dressage teacher, years ago creating a visual 'wall' in front of my horse so that he made a steeper sideways movement in order to reach the track without running into her (second: horses try to avoid running into people if they can). I tell Nikki what I have in mind so that she's prepared (second), and then stand in Ollie's path in my bright red jacket with my arms outstretched to each side, moving backwards and sideways to keep the same distance from him as he moves forwards and sideways. He is watching me intently; his ears are sideways, showing that he is also 'listening' to his rider; his face is calm and his steps are even and rhythmical. He easily manages much steeper angles on both reins. Imagining how this adds up for Ollie, I sense he's (second) calm, intent and almost curious. His body language is saying: *I'm doing the right thing, I think, because they are both saying 'good boy, oh good boy' in that special tone of voice. And the red one leaves me a space to go past her when I reach the track: she doesn't want to stop me, then.* Nikki says that she felt no hesitation in him: she feels (second) that he trusts me, though he doesn't trust everybody. I explore why this might be (third). Thinking back, I was clear in my mind and in my body signals what I wanted him to do, though I felt no sense of urgency or pressure. My energy was calm. I believed he and Nikki could make a more advanced version of the movement, and really wanted to help Ollie discover that he could do more than he thought he could. I was on his side.

Horses can invite – challenge – even force – us out of the comfort of our instinctive first-position experience of the world into the realm of experiencing it differently. I'm sure this happens with everyone involved with them; and it's also a specific goal of various equine-assisted therapies.

> If I could take every corporate executive and put him or her through a few sessions in the round pen with a horse, their understanding of trust would be elevated to such an extent that they would go back into the workplace with a whole new confidence. They would also give more importance to what goes on in the lives of their colleagues, both inside and outside their workplace.
>
> Monty Roberts, *Join-Up, Horse Sense for People*, p.98

Applying what you learn to other situations

'*Learning to wash a horse's mane has encouraged many young people to become independent in their own hair-washing.*' (Fortune Centre of Riding Therapy Website)

The effort we put into trying to understand the horse's experience of any situation is an effort that can also pay off in our relationships with other human beings. Indeed, this is the intention of many programmes involving horses with teenagers or adults whose ability to form and sustain relationships with other people is limited or dysfunctional. It's also the cornerstone of leadership programmes, such as those offered by the English training organizations Acorns2Oaks and LeadChange, which help people to develop their management and influencing skills through working with horses at liberty. Our brains are strongly programmed to make connections – indeed, that's just how learning takes place: little filaments (dendrites) reach out from a nerve cell to link up with filaments reaching out from another, and so new neural pathways for learning and memory are created. Our mental filing systems rely heavily on similarities: logical and structural ones, categorizing ones, feeling ones ('seems like', 'reminds me of') and associations ('our song', 'that time of year'). When we learn something new we internalize its essential structure and pattern, and it becomes potentially available as a template by which to assess other experiences, especially new or unfamiliar ones: if there seems to be a fit, we can transfer what we have learnt and use it in the new situation. We will then, of course, depend on the feedback we get to tell us how appropriate and how useful that transfer has been.

This process of learning transfer can happen consciously or unconsciously. In the case of the teenagers whose horse-care activities prompted them to take more care of themselves, it's very likely that the transfer was unconscious. Probably they had already been told, and reminded, and nagged, many times over the years to wash and tidy themselves more – and probably none of these external messages had translated into a reliable internal programme! It's more likely that making a horse's mane and tail clean, smooth and more beautiful through grooming and washing set off a prompt at a deeper, more feeling level. There's a 'what if...' and 'me too?' element to it. And, essentially, the impulse towards better *self*-care felt as if it *came from themselves*.

Not long ago I spent two days on a LeadChange programme for managers, juggling all three of the perceptual positions discussed earlier! Because it was a training programme, the first-hand experiential sessions with the horses were followed up by reflection and discussion, allowing people to examine how they had felt and what they had done. The horses' clean mirroring allowed individuals to see how their leadership style came across when they were trying to do 'simple' tasks like picking up hooves, or more complex ones like getting horses to follow them on a specified route around the indoor school. People became frustrated, despondent, excited, puzzled or elated, according to the 'results' they got – *and* they also began to understand why colleagues at work responded to that same leadership style of theirs in the way they did. At the end of the course they went away more able to look at themselves from second and third position than when they arrived, and with some different strategies for working with others.

People attend courses for different reasons, and may be more or less open both to the feedback they receive and to connecting it to their lives outside. Even though horses employed as equine assistants in learning programmes are carefully chosen (some therapists would claim that they essentially 'volunteer'), they can still be a genuine force to be reckoned with. It's not so easy to switch off from learning opportunities when your teacher weighs half a ton, can leap many feet sideways in a single movement and accelerate rapidly in an instant! Like the seasonal fires in parts of America and Australia, without which certain kinds of seeds can't crack open and germinate, that stimulus of even everyday encounters with horses can open us up to new growth mentally, physically and emotionally. New knowledge that's acquired in a context of pleasure, excitement, curiosity and even danger is likely to become part of you: it will be 'in your muscle' rather than just 'in your head', and become quickly and reliably your own.

Using feedback to ask bigger questions

By becoming more alert to our own feelings, thoughts and bodily sensations, and more receptive to feedback from others, we can become quicker and more sensitive in *all* our reactions and responses. It's a bit like tuning an instrument: tuning sets it up to play not just more accurately but also at its individual best. The 'conversations' of interaction between rider and horse and between human and human can build on the process of mutual feedback to produce better relationships by doing more of what works (reinforcing behaviour) and less of what doesn't (counteracting or rebalancing behaviour). Both of these responses are experimental, and stand a good chance of improving the situation. Since what you once know cannot again be 'unknown' (only forgotten or ignored), you build a loop-like habit of learning whatever you're doing. See the diagram overleaf.

Even though it can be very efficient, this is a relatively simple pattern

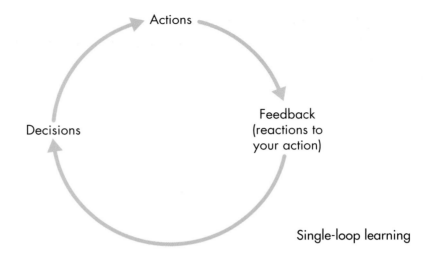

Single-loop learning

of learning. It's been called single-loop learning* because it goes round the same single track however many times you repeat it. Every ride (and every human encounter) you have is an opportunity for single-loop learning as you register the feedback you are getting and react accordingly. (*The concepts of single- and double-loop learning were originally explored as organizational tools by Chris Argyris and Donald Schon in their book *Theory in Practice; Increasing Professional Effectiveness.*)

But we can also go beyond the single-loop mechanism of doing more or less, same or different, and begin to use the feedback we receive as data for asking bigger questions still. We can use the feedback we get to ask questions about the assumptions, norms and patterns that underpin what we are doing. Even if what we're doing works, might it be better if we approached it another way? Even if we've found a solution, could we find ways to generate others which might be better still? What are our underlying principles and assumptions? In taking certain things for granted, have we blinded ourselves to others that might be more important or more useful? We could become experimental not just in a practical sense, but in our very thinking. We could question old rules and begin to devise new ones – not only when we have to because the old rules are proving inadequate, but because we are curious about what they might be ruling out or covering up even when they do seem to be working.

In asking larger questions like these we are adding another, larger loop to our first one and opening up the possibility of using our learning more profoundly and with wider-ranging implications. Double-loop learning looks like this:

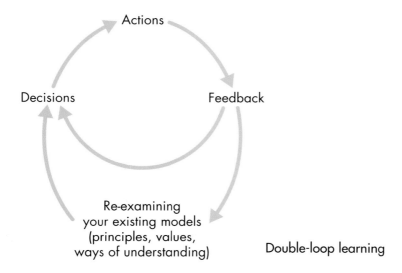

Double-loop learning

The discipline of attending, reflecting and experimenting can thus become truly transformational.

Double-loop learning even gives us a strategy for looking at learning itself. One of the taken-for-granted rules we can look at like this is the rule that tells us we should be paying attention… Not because we shouldn't, but because there's an assumption that we know what 'attention' is. One of the great discoveries that the American tennis coach, Timothy Gallwey, made in the 1970 was that there's attention… And then there's attention…

> The key that unlocked Jack's new backhand – which was really there all the time just waiting to be let out – was that in the instant he stopped trying to change his backhand, he saw it as it was. At first, with the aid of the mirror, he directly *experienced* his backswing. Without thinking or analysing, he increased his awareness of that part of his swing. When the mind is free of any thought or judgment, it is still and acts like a perfect mirror. Then and only then can we know things as they are.
>
> Timothy Gallwey, *The Inner Game of Tennis*, p.31

Paul Belasik's discipleship in Zen brought him to a similar awareness about riding:

> In the best concentration you stop putting things into your mind. You try to cease the directives and analysis; you let the mind be free to do what it does best – and that is to perceive the entire world around you with all your senses at once. Because we are multisensory we can be receiving information on several levels all at the same time. The result of the right concentration can be miraculous. In an open, alert but calm way the body can attain intuitive reactions which are not really reactions because there is no time lag between action and reaction. The horse and rider literally become one, feeling things at the same time.
>
> Paul Belasik, *Riding Towards the Light*, J.A. Allen, p.112

More energy and clearer guidance from me give Mouse the security of knowing what I want, which shows in greater self-carriage and a calmer expression.

The Gift of Presence

The nature of discipline

Discipline is an interesting word. It has come to imply working at something with concentration, which has implications of determination, effort and seriousness of a somewhat dour kind. I think it helps to remind ourselves that *discipline* relates to *disciple*: it's about a master-learner relationship. If we open our minds to the possibility that horses are our teachers if we let them be so, the discipline of exploring that relationship can offer us as much insight, as much delight, as much pleasure, as much fun (and incidentally raise as many of the bigger questions about being together in the world) as riding itself.

Chapter Four

The Privacy of Presence

Horses can teach us to respect our inner feelings without blame and judgement. They can teach us how to be honest.

Ariana Strozzi, *Horse Sense for the Leader Within*, p.89

In the previous chapter I explored how the immediacy of horses' reactions anchors us in the present. In this chapter I want to explore another equally well-attested process: how their very presence can, if we will let it, expand its influence out from the present moment to touch the core of who we are and create space for personal growth and lasting change. And, in particular, I want to consider how horses can help us experience a sense of personal spaciousness that is at the same time emotionally supported. This is what I mean by 'the privacy of presence'.

The notion of 'spaciousness' raises some interesting possibilities. Why is space a good thing? What is space useful *for*? At its most fundamental, all growth requires space. Something that was *this* size grows to become *that* size. If it is hemmed in, it can't do that – or it will be distorted or stunted as its growing edges ram up against the limits of the available space. In physical terms this is obvious, as with plant shoots. But it's also true (if sometimes hidden), in mental or emotional terms. Confining the mind through rigid thinking on the one hand or lack of stimulus on the other has distorting effects. Confining it emotionally has psychological ones.

When I think of physical space, I think of stretching. I think of increasing *ease*. When we moved from our narrow Victorian terrace house to our chalet bungalow in the country, what changed was that each room now gave us enough space to stride rather than taking small steps. Silence and Baroque music offer me space for thought. Having a speculative discussion with my husband Leo or with good friends, or receiving coaching from one of my valued colleagues, means I have space to feel, to explore, to discover and re-create myself. When I am acting as coach it means I am doing my best to offer these same benefits to someone else. The opposites of space in this sense are noise, crowds, haste – each in its own way involving a crowding, cramping-in of senses, mind and spirit.

Horses are space-makers, too. Many people find that being around horses seems to be therapeutic in itself. Just doing the mundane tasks of everyday stable work, grooming and caring has an emotionally spacious, calming effect, as a number of questionnaire responses showed:

'Spending time with horses is a great stress-buster at the end of a busy day. I find peace in just being in their presence and listening to them munching hay or feeling the warmth of their breath.'

'My horse definitely helps me de-stress after a day at work. It is really rewarding when he comes to meet me at the field gate and knowing that his unconditional love will always be there (even if it is just because I'm the one who feeds him!)'

'Very peaceful and mood-enhancing.'

Like coaches, horses give us room while at the same time being attentive. They monitor us without interfering. They may make demands but they do not cramp our style.

These times of domestic interaction with horses seem to allow us to receive from them a sense of acceptance and warmth at a deep level. A number of people also mentioned times of personal trouble (for example, a divorce, a miscarriage, loss of a job) when they had felt soothed and even actively supported by their horse. They felt that their horse was able to sense and respond to the anguish they felt. Therapists who use horses as their partners and assistants in working with troubled people report the same responsiveness and the same sensitive acceptance.

The Gift of Spaciousness

What makes a relationship therapeutic?

Some years ago my husband's job was very stressful: there were days when he really didn't want to go to work, and when the only thing I felt I could do to support him was to walk there with him. At that time, quite coincidentally, the student-groom who had helped at the small livery yard where we kept our horses left: this meant that the routine daily horse-tasks all had to be shared amongst the owners. Since Leo and I were both working full-time, we had to ride our two horses each evening, often feeding, rugging-up, watering and skipping out the others (ten in all) before locking up. Evening meals began at 10 p.m. when we eventually got home! Often, it was hard to gear ourselves up and go out to the yard; yet almost as soon as we got there the horses worked a trans-formation on us, calming us emotionally and re-energizing us again. Their effect was definitely therapeutic: just how did it come about?

I want to look at some of the qualities of therapeutic interaction and to explore how it is that members of another species, relying only on non-verbal means of communication, can help human beings like this. One way to clarify what's involved is to see how horses measure up to what that great therapist, Carl Rogers, founder of what is known as Person-Centred Therapy, identified in a chapter entitled Characteristics of a Helping Relationship. A key point that Rogers makes seems inter-estingly applicable to horses:

> It is the attitudes and feeling of the therapist, rather than his theoretical orientation, which is [*sic*] important. His procedures and techniques are less important than his attitudes. It is also worth noting that it is the way in which his attitudes and procedures are perceived which makes a difference to the client, and that it is this perception which is crucial.

> Carl Rogers, *On Becoming a Person*, p.44

Rogers asked himself what qualities and behaviour a therapist had to demonstrate in order to create a truly helping relationship: in reminding myself of what he said I was struck by how many of the conditions he identified as 'necessary and sufficient' can also be demonstrated by horses in their relationships with the people around them. I extrapolated the following from his work, just cited.

Mouse's dry hooves need daily moisturizing: it's a 'mindless' routine which brings a small sense of achievement and a few minutes of calm emotional space.

- Show a way of *being* which will be perceived by the other person as trustworthy, dependable or consistent in some deep sense.

- Communicate unambiguously.

- Show positive attitudes such as warmth, caring, liking, interest, respect.

- Have a clear identity of their own.

- Be secure enough within themselves to permit others their own separateness.

- Show empathy with the feelings of others.

- Accept others for who and what they are, without being judgemental.

- Be sufficiently sensitive in relating to others not to appear a threat.

Each of these qualities contributes to the making and maintaining of good relationships within the equine herd – and at the same time sends

The Gift of Spaciousness

a powerful message to human beings: 'I know who I am; I get on with who I am; I accept you for who you are.' If we bear in mind that the effectiveness of therapy depends on how the *client* perceives the therapeutic relationship, then, by implication, a horse who gives out messages like these will be offering support and space to his human carer that's potentially as powerfully therapeutic as that offered by a trained 'person-centred therapist'.

It has been argued that the essence of the relationship created by having this kind of attitude towards the other person, is that:

> We come to know and value and respect the client not because he is good, exemplary, or what we ourselves would like to be, but because we understand him and his life experiences from his own internal frame of reference rather than from an external 'objective' viewpoint.

Charles B. Truax and Robert R. Carkhuff, *Toward Effective Counselling and Psychotherapy: Training and Practice*, p.42

Extensive reviews of therapeutic literature further boiled down the essence of relationships that facilitated change to three essential qualities: *genuineness, accurate empathy* and *non-possessive warmth*. There's no doubt that horses are genuine (as someone said, even when they're devious they are so in a direct and honest way!), or that they are often *warm* in an emotional as well as physical sense. *Empathy* is less easy to demonstrate. Yet for a horse to have an accepted place within the herd he has to develop a considerable degree of sensitivity towards his fellow members. Indeed, from her observations of the horses in her naturally managed herd, Marthe Kiley-Worthington believes that this awareness extends to the feelings as well as the actions of others. In *Horse Watch, What it is to be Equine* (p.355) she argues that horses have evolved as 'good natural psychologists', adding that 'The simplest immediate explanation for reciprocal altruism is awareness of others and their individual needs and likely response. This means, of course, that they are aware of the others' mental states.'

What is her evidence for believing this, and why does she believe that horses in the wild need develop social skills?

The Gift of Spaciousness

In order to be accepted by other equines, they have to learn the rules of equine society, in other words they have to become good 'natural psychologists'.

The social knowledge that equines have to acquire during their lifetime is considerable. They recognize individuals (in some cases large numbers), and must be able to predict their behaviour. They develop different roles within the society which are moulded by their individual ages, personalities and knowledge of the environment and others.

…there are strict rules of conduct in this equine society which are based on fostering cohesion in the group rather than competition. In addition, each animal behaves differently and the only way to accurately describe them is by individual personality profiles. All of this indicates that, just as in human societies, the organization of this equine society was more complex, flexible, adaptable, tolerant and cohesive than has previously been thought…

Habits form within the society. One of these, which is striking, is to allow the individual social liberty within the group. Another is to develop a great awareness of the environment within which they live. Finally, their social behaviour is most commonly structured to keeping the group together.

Marthe Kiley-Worthington, Horse Watch, What it is to be Equine, pp.271–3

What Kiley-Worthington is talking about is a mutually given-and-taken freedom to move within a *social* space.

I think it is important that, in our gratitude for what horses do for us (and what they do can indeed be profound), we are mindful that their attitudes and behaviour have valid reasons in *equine terms*. There seems to be an interesting and substantial overlap between what works for horses in their own social world and what is beneficial for human beings: it's as though we can benefit – almost by the way – from behaviour that wild horses evolved as a survival mechanism. This is not to deny that, over time, horses become close to specific human individuals or groups: they clearly do. But we need to be mindful that the meaning this has for them, in their world, is not quite the same as the meaning it may carry for us in ours.

The Gift of Spaciousness

This mare has a special talent for therapeutic work. On this occasion, she is acknowledging Rio, who is not a 'client' but a fellow-professional. There's a real sense of appreciation and affection in both their faces.

For example, as mentioned in the previous chapter, therapists who work with horses often say that their equine partners virtually 'volunteer' for the work. And, indeed, some therapists explicitly depend on these horses, at liberty in the arena, 'seeking out' or 'choosing to work with' particular individuals amongst those who have come together for help. Rather than seeing this as some kind of mystical or nurturing process, I think it makes more sense to see such horses as being those who have a particular talent for social awareness. This process of understanding about people simply by observing them has also been described by John Cleese and his therapist, Robin Skynner, in their book *Families and How to Survive Them*. Skynner is talking about a training exercise for therapists, in which:

> They're put together in a group and asked to choose another person from the group who either makes them think of someone in their family or, alternatively, gives them the feeling that they would have filled a 'gap' in their family… The astonishing thing is that… just by *looking*, they choose people who have astonishing similarities in childhood experiences and specific family problems, too.
>
> John Cleese and Robin Skynner, *Families and How to Survive Them*,
> pp.17 and 19

The Gift of Spaciousness

When we *look*, it is evident that we can *see* 'far more than meets the eye'. Within a group of horses there are some individuals whose sight or hearing are particularly sharp, and whose speed and subtlety of perception will help the herd function better: it seems to me that there will be also others who have highly developed 'social skills', and that these are also likely to be useful to the group in helping it to function and, indeed, survive. A herd in which relationships are relatively calm, normally conflict-free and mutually helpful in terms of special skills (hearing, sight, assertive and affiliative leadership roles) is more likely to survive and thrive. In *Horse Watch, What it is to be Equine*, Marthe Kiley-Worthington comments: 'The way equines stay together is because they foster the cohesion of the group: they are 'stickers' not 'splitters'. In other words, instead of competing, they demonstrate co-operation (affiliation), and often ignore threats and disruptive behaviour.'

It would seem that individual equines have some inbuilt programming to serve the needs and well-being of the herd, and this may also be true of those with acute social sensitivity. I have sometimes wondered whether we are justified in 'using' horses as helpers in therapeutic situations, but anecdotal evidence seems to suggest that they are willing 'volunteers'. It does occur to me on reflection that such horses are contributing their special social awareness to an adopted, if temporary, 'human herd' (often led by their human owner), in much the same way that they might contribute it within a herd of equines. If this is so, then they are simply transferring a particular, innate behaviour from one group situation to another – and we are the beneficiaries.

Horses and therapeutic space

Horses at liberty are always on the move. Yet, as Kiley-Worthington observed, they seem to allow each other room to manoeuvre. Even when grazing, a horse isn't still for long, and even though he may keep close to his particular friends there will be a varied scattering and shifting amongst the group members even in a single field. The patterns are those of independent movement within an ongoing, often complex, web of connectedness. The web extends across the shifting physical spaces between the members of the group. When they are free from fear,

The Gift of Spaciousness

Although they have the freedom of a huge field, these two horses have clearly chosen to place themselves in relation to each other. They both have access to the shade if they want it, and by facing in opposite directions they can support and protect each other if need arises.

conflict or hunger, horses can simply *be*, and allow each other to be. It's as though this is a natural, 'default' state, to which they return when any particular need or minor emergency is over. It's akin to the recent emphasis on 'contentment' rather than 'happiness' as a desideratum for human beings. In his book *Enough: Breaking Free from the World of More*, John Naish quotes Shen Sh'iàn, the editor of *The Daily Enlightenment* newsletter, as saying: '*The moment we are content, we have enough. The problem is that we think the other way round: that we will be content only when we have enough.*'

This, too, may be one of the ways in which horses model behaviour which has helped them adapt to their environment – and which, if we allow ourselves, they can lead us toward in our turn.

When horses are in this state of quiet contentment, as when they're munching quietly, allowing us to groom them or to carry on everyday tasks around them, some important things happen.

Physical space and its far-reaching effects

At an unconscious level we can begin to pick up the physical rhythms of horses' breathing and movement (both, in these undisturbed situations, slower than our own) and, as a result, be led into a calmer mind-body state. This is the mirror of the same process by which horses often 'resonate' with the emotional states of human beings:

> Recent work by Candace Pert and other researchers active in the field of psychoneuroimmunology shows that the molecules carrying emotional information (called neuropeptides) are not only generated by the brain, but by sites throughout the body, most dramatically in the heart and gut. When people have 'gut feelings', they're not speaking metaphori-cally. As animals possessing extremely large and sensitive guts – and hearts for that matter – horses have huge resonant surfaces for receiving and responding to emotional information… [and, by implication, generating it].

> Linda Kohanov, *Riding between the Worlds*, p.xvi

When we enter into a horse's space with the intent of caring for him or simply enjoying being with him, we place ourselves in a position where it is easy for us to be influenced, or 'led', by him. He takes between 8 and 16 breaths per minute: we take between 12 and 18. His heart beats between 32 and 44 times per minute: ours beats on average 70 (woman) and 75 (men) times. 'On his turf', his naturally slower biorhythms are more likely to influence us to match him than the other way about.

Grooming is particularly interesting, because it seems likely that it has a reciprocal calming effect. It has been found that: 'when tame horses were groomed by humans at the site they most frequently chose when grooming one another (the lower neck), their heart rate decreased significantly – about 11 per cent in adults'. (Reported by Stephen Budiansky in *The Nature of Horses*, p.88.)

By implication, then, it seems that when humans groom horses this is likely to initiate a mutually beneficial circuit of calming: as we groom the horse his heart rate slows; as it does so, it influences us so that ours begins to do the same. In Chapter 9 I shall refer again to such changes in mind-body states and explore more fully how, over time, they can intensify and enrich our whole experience of being alive.

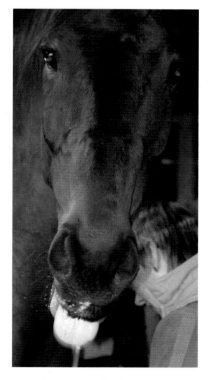

Though Mouse is looking at Leo, his serene presence is clearly having an influence on me.

The Gift of Spaciousness

Spacious connectedness and personal learning

When we allow ourselves to join with horses in spacious connectedness, rather than urgent task-determined activity, we can enter into a state of emotional and mental openness. We may become more aware of our thoughts, feelings, or physical experience. We may find our minds free-floating or making unfamiliar connections. Through the very concentration at one level on what we are doing, we can become more open at other levels. We are in a state, not of certainty and constraint, but of *flux* and *possibility*.

I think this is close to how the psychologist D.W. Winnicott described *playing*. 'Playing facilitates growth and health; playing leads into group relationships; playing can be a form of communication in psychotherapy… The natural thing is playing…' (D.W. Winnicott, *Playing and Reality*, p.41.)

I believe that when we are involved with horses in this mutually aware, permissive, open connectedness, the therapeutic possibilities of play become available to us. It can happen, of course, when we are actually playing together, as in something like loose-schooling, jumping or relaxed hacking; but it can also happen in any shared activity that does not feel driven, regimented or anxious. Where there is shared activity without regimentation, there is the possibility of learning and growth. As Winnicott says: 'Psychotherapy takes place in the overlap of two areas of playing, that of the patient and that of the therapist. Psychotherapy has to do with two people playing together.' (Ibid., p.38.)

Winnicott's description of psychotherapy as overlapping play was quite revolutionary – and hardly taken up, even at the time; yet in my experience it gives us one of the very best descriptions of all the permissive, non-judgemental, person-centred forms of helping. When I am coaching, it often feels like play – for play involves serious engagement in the moment and commitment to the process and to where it leads. It's playful even if the content is important – perhaps *especially* when the content is important – and even when both coach and coached are serious in their commitment to the process and to one another. Play is about involvement, exploration, awareness, delight and discovery. Because it is so much about what's *possible*, it engages us in a very profound and liberating way. Therapy can be playful even when it

concerns sadness or tragedy. I remember years ago ending a 'heavy' session with a client; as we went down the hallway he turned to me and said, 'Wasn't that exciting!'

I think that horses offer us the chance of playing in this important sense. Our play overlaps with their play, and as a result we can become more grounded in ourselves and more able to navigate through our own complex worlds.

One respondent to my questionnaire put it like this:

I really don't know how to explain it. I know that it is very simple and intricately complex at the same time. They make me feel better. If I was upset about something as a kid my mum used to send me out for a ride and I always came back feeling better. They help me connect with nature and help me regain a balance and sense of how I fit with it. Their honesty, generosity and spirit inspire me and their quiet serenity touches my heart.

This person shows very clearly how horses can, at one and the same time, comfort us, 'take us out of ourselves', enhance our sense of who we are and connect us to a wider context that is quite literally 'grounding'.

This picture catches one of those moments when play in Winnicott's sense is about to happen: there's a connection between Aurele, the human therapist, her horse and the two women which is close, expectant and full of possibility. This is both a peaceful and a joyful space.

The Gift of Spaciousness

Room to be ourselves

In their non-judgemental yet attentive way of being with us, horses
create the supportive space in which we can discover that we can experience ourselves without judgement but with honest self-awareness,
coupled with the possibility of learning and change. In his horse-assisted
therapy programmes for teenagers and adults, the American trainer
Wyatt Webb explains how this can work:

> I explained to these kids that everything they knew regarding how to
> treat other living things was primarily learned from people. Bottom line:
> I knew that they'd treat the *horses* just like they'd been conditioned and
> taught to treat *people*. Behaviorally, the kids in the program had the
> opportunity, through their interaction with the horses, to look at what
> they'd learned: was their behavior working for or against them?

Wyatt Webb with Cindy Pearlman, It's Not about the Horse, p.80

Webb goes on to say:

> …when humans begin to work with horses, what tends to happen is that
> things don't always go smoothly, so we tend to make up a story about it.
> In other words, we *diagnose* the problem. And once we've done that,
> we'll usually make it about the horse… *let's not do that anymore.*

Ibid., p.140

The point that I want to make is that, by co-creating a space in which
each of us can be private yet present with the other, horses come close to
offering us what the best human relationships can do. Where people's
life-experience has been limiting, frustrating or damaging, experiencing
different possibilities via equine-assisted therapy can both heal old
wounds and teach new skills for living. But I'm sure that the same shared
privacy of presence can enhance lives that by and large feel good enough
already, taking those lucky enough to live them closer to that ideal of
contentment which every so often slips over into outright happiness.

I am standing with my friend Nikki's horse, Merlin, in his box. We
have known each other for eleven years, since he was a 2-year-old. He is

intelligent, friendly and curious, but I don't think there's any special
bond between us. On this occasion I am waiting for something or someone, and stroking Merlin's neck from time to time. Something makes
me decide to blow, very gently, in his ear. He becomes very still – it's as
though he is considering what's just happened. Then, very slowly, very
deliberately, he turns his head so that his neck makes an arc in front of
me – and blows in my ear.

The Gift of Spaciousness

Chapter Five

The Place in the Herd

Humans did not invent leadership. It exists throughout the animal kingdom as a practical way for social animals to communicate, negotiate, and collectively contribute to the survival of the individual and the whole. As soon as two social beings begin to relate, a powerful but subtle request for information is made: *Who is leading? What are we doing together? What are we communicating? What needs to be negotiated? What is my role? What is your role?*

Ariana Strozzi, *Horse Sense for the Leader Within*, p.45

Horsy people quite frequently say that if a rider/handler approaches a horse in a similar way to the alpha stallion or mare in a herd, it will be easier to elicit the horse's co-operation, because this kind of behaviour makes sense to him in his own terms. The rider/handler will be taking up the obvious lead-role, so the horse will do what he is being asked. It's tough, of course, for those of us who aren't alphas by nature or who happen to find the idea of managing with an implicit threat of teeth and hooves distasteful. While this model makes us consider how horses might understand our behaviour in their terms, to be truly useful it needs more fleshing out and further examination. In this chapter, I want to explore what herd behaviour and herd roles can really teach us about influence, and to suggest a variety of ways in which an understanding of the different means and styles of influence that operate amongst horses can help us in our interactions, not just with them, but also with other human beings.

There are many ways of influencing others: in what direction would you like to lead them – and what kind of leader do you have it in you to be?

Reflections on influence

1. I am judging a large Medium dressage competition, a qualifier for the Summer Regionals, in which there are some stylish riders and some enviably talented horses. I am tossing 7s ('fairly good') at their sheets like confetti, and there are a few 8s ('good') as well. What really thrills me, though, is not one of the spectacularly talented horses or 'look-at-me' riders but a good average horse with his perfectly attuned rider. They are more than accurate: they are thinking the same thing at the same moment – and it just happens to be at the specified marker. They swing softly in harmony to each other's movement: everything about them is elastic; nothing is stiff. You can't tell who is giving and who is taking – it's an ongoing conversation which seems as friendly and low-key as if, rather than describing patterns in an indoor school, they were hacking through the fields on a sunny day. Of course, the rider is directing their actual route, but they are enjoying it *together*. The horse's rhythmical onward movement, his swinging tail and his attentive ears are all clearly saying: 'OK, Mum, where next?' In the end, they win the class, beating a more extravagantly talented horse by less than one per cent. Their 7s were a consistent baseline above which the 8s I gave for his obedience and her riding stood out as defining (and doubly weighted) high points. There wasn't much between the first and second pairs – only the *exceptional quality of the relationship*.

2. I have started coaching the head of a business department whose staff and remit are world-wide. As part of our programme, I interview some of her immediate team. She generates huge loyalty: 'We'd die for her', one team member says quite straightforwardly. They tell me how loyal she is to all of them: she is direct and can be blunt but she is also

realistically appreciative and immensely encouraging. Yet I don't get a sense that she's unduly protective or defensive. As she tells me about negotiating some tricky situations involving both colleagues within the firm and outside contractors, I realize that, while she is highly strategic in her approach and keenly aware of 'setting up' situations to facilitate the outcomes she's after, *she manages to leave everyone she deals with better off*. Management jargon calls this 'creating win-win'; but I'm well aware that people can tell when someone is manipulatively buttering them up or giving them a quid pro quo and when, by contrast, there's a genuine exchange based on a real wish to further everyone's well-being. This manager leaves a wake of good feeling and well-being behind her everywhere.

3. I am riding Mouse in the school and reflecting on what it means for him, and for me, to be soft in the back. When we got Mouse, our aim was to help him relax and use his back more fully, and in so doing to discover – as it were by accident – that rounding up onto the bit was something that could come naturally to him without any coercion or constraint. Coincidentally (or was it?) I have had a long-lasting back problem of my own, which has only responded slowly to various forms of treatment, so that though I didn't actually stop riding, even when it was most acute, I have, for over a year, been enduring stiffness, recurrent feelings (and sounds!) of crunching and the pain of a pinched nerve.

    There's a 'back-story' to our work in the school today. Nowadays Mouse is generally softer and more 'through' in canter, and so probably am I, which was why at one point I decided to concentrate on canter for a while and hope that we could capitalize on any softness and roundness we achieved there and carry it over gradually into little episodes of trot. However, I lost this focus one day and, because of my frustration, ended up both doing more trot and being more demanding about rounding up. The result was that Mouse and I resisted each other and I got sore shoulders! (I'm not sure what he got physically, but he wasn't happy.)

    Angry with myself for having compromised what I'd believed and intended, and feeling that I'd set our progress back, last time I rode I wasn't really sure what to do for the best. We started with canter, but

Mouse and I were both stiff and I was sore. It all seemed too much, and I have since been feeling discouraged, both about him and about the non-improvement of my back. Then I remembered a lesson with my classical trainer, Ali, when she'd suggested I ride without stirrups in order to be able to wrap Mouse more closely with my legs and so encourage him to lift his back. Why not try that? So today I'm riding without stirrups.

My legs seem hugely longer and feel much closer and friendlier to Mouse's sides. He starts rounding up for longer bursts, though still not very consistently. I think: *If I want him to give me his back, I must give him mine.* I think fleetingly of the additional pain that tucking my pelvis under and really swinging with him will probably cause, but at this moment it seems more important to be giving to him than protecting myself, so I do it anyway. Actually, it's much, much better. *We're* much, much better. We spend about twenty minutes in our slow, rhythmical trot, relaxed and peaceful as we do big shapes and soft turns together. That seems like enough, and good enough.

As the rest of that day goes by I realize that the pinched nerve in my back is now free. I can't quite believe it. My back feels weary, but it's a coherent, connected back again instead of a collection of jarring bones and overstretched ligaments. Mouse and I have given each other the freedom of our backs, in the process of which he has done something quite fundamental to how my back feels and works. The positive effect is like re-choreographing a dance to make it smoother, more flowing, more effortlessly expressive. I read in an article about riding for the disabled that this is something horses can do for people with physical disability – they can actually effect changes to the neurology of the spine. That's how this feels to me today.

What models of herd leadership do horses offer us?

In recent years the thinking about how humans can best influence horses has drawn upon observations of how horses influence each other in the herd. There are two obvious 'leadership roles' within a herd: the mare, who socializes young herd members and who, in times of danger, selects the best direction in which to flee, and the stallion, who selects and

The Gift of Influence

mates with his small harem, keeps his herd together, defends them and, when necessary, drives them towards safety. These two alpha individuals lead, shape and define the harem group, and in larger herds they have to compete and negotiate with others in comparable roles. In terms of our interaction with horses, both offer us an example of clear communication, but the way they interact is limited by being essentially dominant and socially coercive.

Nonetheless, this is a model that many people seek to emulate, both in training horses and in management circles. Even Monty Roberts' round-pen method of sending the horse away until he seeks 'join-up' has an element of this, deriving as it does from the social disciplining used by alpha mares to socialize unruly youngsters into the ways of the herd. Roberts' charismatic breakaway from the rough, dominant and often cruel methods used by old-school horse trainers (his father amongst them) has led many people to seek a more harmonious way of teaching the horse – and other human beings – what is appropriate; but some therapeutically orientated writers (Ariana Strozzi among them) have pointed out that the model of 'join-up' still, to an extent, supports a hierarchical relationship between human and horse rather than one based on what my NLP colleague Jan Pye and I call 'an equality of difference'.

When he was a teenager, Monty Roberts observed how mares taught youngsters the rules of acceptable herd behaviour:

> Press the young horse away, and his instinct is to return… It was as if the synapses in my brain all clicked at the same time to tell me I'd found what I was looking for. A name sprang to my head – 'Advance and Retreat'… Perhaps I could use the same silent system of communication myself, as I'd observed employed by the dominant mare. If I understood how to do it, I could effectively cross over the boundary between man – the ultimate fight animal – and horse, the flight animal. Using their language, their system of communication, I could create a strong bond of trust. I would achieve cross-species communication.

Monty Roberts, *The Man who Listens to Horses*, pp.103–4

Immensely detailed observation of the lead-mare's powerfully educative role eventually enabled Roberts to learn, then teach, the 'language' he

Katie's influence on her foal is clear not just from their closeness but from the way the foal is copying his mother's stance and the orientation of her ears. He is even synchronizing his leg movements with hers.

calls *Equus*, following the repeating pattern of advance and retreat with young or difficult horses in the round pen until 'join-up' is achieved. For him, and for many horse-lovers, this has produced amazing results and much happier relationships between humans and horses.

Yet it is only a part of the picture. Roberts' training methods are largely based on the motivational influence generated by the mare (modelling valued behaviour, regulating inclusion and exclusion), and those of the stallion (drive, group cohesion and fear). These are essentially what NLP describes as '*away-from*' motivations. Some human beings, and perhaps some horses, both understand this and feel secure within it. What's important, though, is that within the natural circumstances of the herd, the artificially created circumstances of the round pen and the 'be charismatic and tell-'em' type of leadership manual, 'away-from' behaviour is usually most appropriate in emergency situations rather than being a good model for all situations.

The opposite approach is to generate a '*towards*' motivation, to develop what Robert Dilts, one of the co-developers of NLP, enshrined in the title of a management book he wrote called *Visionary Leadership Skills: Creating a World to Which People Want to Belong*.

In *Horses Never Lie*, writing about the same time as Roberts and just a few years later than Dilts, Mark Rashid expresses his own sense of reservation about following the alpha model – and it's a reservation that's also drawn from long-term observation of horses:

> I needed to truly understand how he [my horse] sees his relationship with the alpha of the herd. Is that a relationship he likes and is happy with? Is the alpha a horse in the herd that he looks up to, respects and follows willingly? Or is it a horse that he would rather not be around in the first place?

Mark Rashid, Horses Never Lie, p.21

At our yard, one horse is so aggressive in his behaviour (so stallion-like even though he is a gelding) that he can only be turned out on his own. If he is with others he attacks or drives them, making them anxious and unhappy. He's not a horse they'd choose to be with!

In addition to the leadership provided by alpha mare and stallion, each herd will rely on particular individuals for specific qualities and skills they offer: horses with good sight or hearing, for example, will contribute to the safety of the herd through the acuity and speed of their reactions to signs of possible danger. But there is still another way of contributing to the well-being and cohesiveness of the herd. Rashid points out that, through their evolution as prey animals, horses are programmed to conserve their energy in case it is needed for life-saving flight, and continues:

> Primarily they conserve energy in a herd situation by willingly following a leader that they know won't cause them unnecessary stress or aggravation. In the herds that I had a chance to work with, it was evident that seldom, if ever, was the chosen leader the alpha horse. Rather, it was a

horse that had proven its leadership qualities in a quiet and consistent manner from one day to the next. In other words, it was a horse that led by example, not by force.

Ibid, p.38

For want of a better term, Rashid calls such individuals 'passive' leaders: their leadership role is acquired as the result of being chosen by others. Their behaviour gets them, and their affiliates, what they need and want – but peacefully and often indirectly. Rashid describes a number of inter-actions which, in his view, demonstrate such horses' indirect, strategic way of goal-seeking, and an influence that, even though it is clear, is non-confrontational. As far as we know, horses don't think strategically in the long term, as humans do, but it's clear that they connect cause and effect to sufficient extent that they exercise some degree of choice in the moment. It may also be that the horses Rashid sees as 'strategic' are ones who have the intelligence and awareness to find out what works without being directly confrontational.

On the day Mouse arrived at our yard, we put him in his new box in the open barn, next to a horse called Slim who had been at the yard for some time. Slim was curious, repeatedly sniffing and mouthing Mouse over the low wooden barrier between their boxes. After a while, Mouse simply turned his back and rested one hind leg: he didn't kick out or bite – but that hind leg clearly said something like: *'Enough already!'* In the field hierarchy, Mouse comes second, after Gordon but before Slim: when someone goes to the field to bring them in, that is the order in which they stand waiting.

The job of the alpha animals in the herd is to lead it in specific situa-tions of danger and to pass on healthy genes to the next generation. 'Passive' or chosen leaders, rather than the alpha horses, seem to be the ones who actually do the day-to-day non-emergency leading within the herd. They motivate others to follow them because being in their pres-ence offers strategic advantages, individual protection and emotional affiliation. As Dilts has expressed it: 'Leaders do not communicate through bragging [the stallion], or threatening or criticizing [the mare], but rather through what they are able to achieve through their skill and their vision.' Which brings each of us to a key question.

The Gift of Influence

What kind of leader do I want to be?

*It seems to me that if we want to emulate horse behavior during training, we must decide what type of horse we are trying to be like. Should we try to be like... the boss horses that others seemed to respond to out of fear and distrust? Or do we want to be like... the horse that can find a quiet solution to even the most difficult problems – the leader that other horses **want** to be around and look up to?*

Mark Rashid, *Horses Never Lie*, p.62

As horses mirror us back to ourselves, they give us the chance to understand how our own attempts to influence others actually come across. This process is at the heart of leadership training courses that involve the participants working with horses running free in the schooling arena. Exercises such as picking up horses' feet, attempting to lunge them, leading them without pulling on the lead-rope (or, indeed, without a lead-rope), getting them to go between obstacles or into a marked-off space, all raise the question of *how* participants are going to make their wishes clear and persuade the horses to go along with them.

One course my husband Leo and I went on gave us the opportunity of discovering in a very special way just how much is involved in allowing yourself to be led. Participants were paired up, and one partner from each pair was blindfolded. Then the seeing course members led their blind partners around an arena, in which a number of obstacles had been built. The seeing person's task was to guide their partner around, over and through as many obstacles as possible, using only non-verbal signals. I was blindfolded, and did my best to follow and trust my partner, while Leo, who was not blindfolded, had the opposite task with his. I became very aware not only of my partner's hand-pressure but also of the energy and degree of muscle-tension being conveyed through his arm. I was trying hard to be a 'good' partner and to do what I sensed was being asked of me. After a while Dave, the course leader, asked us to stop for a moment, and told the seeing people to switch partners and lead someone new. With my new partner I felt an extra sense of comfort and courage, so much so that when I sensed him (probably a male, I

thought) wanting to go faster I was comfortable enough and trusting enough to take the risk of running freely forward alongside him. I marvelled at myself being willing to abandon caution and do this – and it actually felt joyously liberating. Only when the blindfolds were removed did I hear laughter and realize that my 'new' partner was Leo – to whom I'd been married for over thirty years! I had had no idea – but I *had* felt that I could trust and go with him, wherever he led, with confidence and even joy in risk-taking.

None of us can be successful in leading others if we try to make ourselves lead in ways that don't feel appropriate to us: our energies will stutter or be blocked; we will lack the clarity of vision and intent that could otherwise come across as 'meaning it'. Also, it's very likely that in the effort to try to be someone we're not we will lose the ability both to empathize (take second position) with those whom we wish to influence, and to be able to judge what is happening and whether or not we are being effective (third position).

I think it makes most sense when we go back to the horses again: if we can learn to recognize what our special abilities are, and what role we are being invited by others to take in relation to them; if we can be true to our sense of self and our deepest beliefs about what we are trying to

Just what does it take to influence a horse? Most non-horsy participants on equine therapy or leadership courses begin by trying to 'explain' what's wanted – in this case by boxing the mare in so she had to 'jump' out.

The Gift of Influence

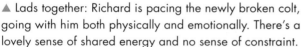

▲ Lads together: Richard is pacing the newly broken colt, going with him both physically and emotionally. There's a lovely sense of shared energy and no sense of constraint.

▶ Only a little later in the same session, the lads have become young men. There's a new dignity and poise in their carriage, yet Richard's hands are still just as open and allowing and the colt clearly feels he is carrying himself as well as his rider.

◀ With Gaz to set him an example, the youngster has become a grown-up in his stance and behaviour. He knows how handsome and well-behaved he is, and he wants his audience to appreciate it too. Though Gaz has a stick and the youngster has a chain on his bridle, the real influence here is that of attraction, not coercion.

The Gift of Influence

do and how we are going about it; if we can be sensitive to the feedback we are getting from others and flexible enough to adjust our behaviour when they're telling us we have lost them, failed to convince them, or made them uncomfortable – then we have created the kind of relationship in which they will actively seek out our company and our influence. In organization-management terms:

> Purposes are the products of possibilities and potentials. You will only be able to spot the possibilities if you understand the external system, and you will only be able to help the organization realize its potential if you understand the internal system and the nature of your influence within it.

> Anne Deering, Robert Dilts and Julian Russell, *Alpha Leadership*, p.154

In other words, it is more helpful, and more realistic, to think of 'leadership', not as a force exerted downwards, but instead as the attractive power that invites one member of a community to join forces willingly with another within a matrix of links and possibilities. This holds true both in my place as a human seeking partnership with my horse, and as a human being seeking to be myself and use my unique skills in harmony with family, friends, colleagues, clients, and strangers…

Finding new ways to be influential

The writers Paul Belasik and Mark Rashid both describe from their own experience how familiar, and even successful, leadership styles can become rigid and limiting, around horses and in life generally. Indeed, through the discipline of martial arts, both of them discovered how pointless it was in the long run to try to achieve mastery in a specific area of their skills if they were not willing to see how their initial approach (in martial arts, just as in riding), expressed the way they lived life as a whole. Making this mental jump allowed both of them to become receptive not only to the 'way' of the martial arts but to their horses and to themselves. *Influence over others begins with self-influence.*

The Gift of Influence

I saw…the struggle within myself against a system suggesting and supporting power, force, domination and aggression in a man. My job,
even when I didn't know it, often involved learning not to be afraid of
softness; being comfortable in allowing softness and intuition a place to
thrive in the sphere that was me. I had not only to let it exist, but also,
like the great Japanese wrestlers who study gentle painting, I too had to
cultivate the opposites in order to develop my balance.

Paul Belasik, *Riding Towards the Light*, p.123

This is one of the themes of equine-assisted leadership courses. The
experience of working with horses at liberty offers people a chance to
discover how their habitual style of influence comes across to others, and
to investigate alternatives that will enrich their personal 'repertoire' and
help them become more effective and at the same time more congruent
with themselves and with others.

The horse teaches the bossy leader to learn a new way to lead – through
encouragement and inclusion. The cheerleading leader learns from
horses to have constructive boundaries and set clear conditions of satisfaction for the team. The appeasing leader learns how to take a stand and
direct her team with intention and assertiveness. The coach learns to
guide clients as opposed to following them around. The pull-along
leader learns to delegate and trust their team.

Ariana Strozzi, *Horse Sense for the Leader Within*, pp.45–6

When I first read this paragraph, I winced. One of the skills that makes
me a good coach is my ability to pace… pace… pace… my clients. I am
actually quite good at moving from pacing them to leading them – the
indirect kind of leading that first elicits from them their inner knowledge
about where they need to go and then guides them into finding and
taking the steps that help them take that desired direction in reality. As is
so often the case, though, a characteristic strength in one context can, in
another, become a weakness: I know I could go on pacing a horse forever
without him choosing to go where he needs to – because going somewhere else, or doing something else, *is not part of his inbuilt agenda.*
Unlike my human clients, he doesn't engage in introspection and has no

Because she is clear and congruent in herself, this course participant can lead two horses effortlessly. The three move as one, and the horses' tranquil faces tell us how comfortable they are following this stranger's leadership.

deep sense of his as yet unrealized possibilities. If I want him to change, I have to show him the way.

So far, so good. I agree with myself completely! But this is where my difficulties begin. I am not a directive or bossy leader – and I don't want to be. I rebel at the idea of 'making' anyone else do something – even if it's in their own best interests. When I try to 'tell a horse what to do' it doesn't convince him and it distresses me – because it 'isn't me'. I know that it helps a horse in the long-term to lift his back, round through his top-line and allow himself to accept the contact of my hands and of the bridle – but how can I explain this to him in a way that is clear to him without doing violence to my sense of who I am?

The inner conflict I experience between doing what I know to be useful and what I feel comfortable with means that I am slow to 'ask' Mouse to round up, and when he does I immediately 'reward' him by dropping the contact as I give the reins forward again! Nikki and Ali, my patient mentors, describe this rather differently. Ali urges me to 'keep supporting him – don't abandon him by yielding the rein as soon as he has taken it'. Nikki points out that, rather than 'giving' the rein I 'throw it at him', and that 'he doesn't like you fiddling with his mouth'. They are both right.

As a rider, I feel like Cinderella: I want to go to the classical riding

The Gift of Influence

ball, but I lack the magic wand that will transform my pumpkin thinking – and turn my personal Mouse into an elegant dressage horse.

This morning I have been for a restful hack around the farm with my friend Gaz. We have been talking about energy, about the molecules that make up mind-body information systems, about how animals understand human energy and intention because they sense this information *from the way we are, not just what we do*. I decide to go down to the school after hacking and to follow up the trot work without stirrups that, only a few days ago, helped Mouse come 'through' and massaged my back into renewed wellness. Gaz stops and dismounts by the barn where his horse and Mouse both live. I am going on past. Except that Mouse is not. No way. He stops, spins round, then, as I turn him towards the school again, he bounces in alternate threats to buck and rear. Gaz strolls up behind us – not close but present. 'Get him to stand facing the school a moment. Now ask him firmly with your legs.' Mouse moves forward. 'Go on asking – every step – *and mean it!*'

Down the track to the school we go. It's a couple of hundred yards – quite a lot of steps and quite a lot of 'meaning it'. No problem. We go in, we shut the gate. Now we are having a different kind of conversation. Clarity works – the heels were only a token expression of it. The clarity that really makes the difference is in my head: it's the clarity of *meaning it*. Just as horses responded to the business leaders on Ariana Strozzi's courses, Mouse responded to my fear of what his bouncing might lead on to and my long-term doubt that I had enough skill as a horsewoman to get him to do what I wanted. And then, when 'meaning it' became a reality in the whole of me, of course he responded to that instead. Though the normal pattern in Mouse's world is that going for a hack ends with a return to his stable, my really, truly *meaning* that we could do something extra was enough to convince him.

And of course this carries on. Once in the school, our bodies remember from last time that we can connect through our backs, so he is able to swing and move, and feel confident enough to find more energy. And 'meaning it' helps me give him a consistent encouragement with my legs and a more constant support with my hands. With this clarity and support *of course* he can carry himself and move in an outline. He can even understand what I want when I ask for travers.

Twenty minutes later, it feels like time to stop. But I have another idea – I want to confirm the new understanding that we have just achieved. I want him to go past the barn again and walk the other way, as though we were leaving the farm again. Despite my intending thoughts and reminding heels, Mouse veers firmly towards the barn, then into it. I stop him. He bounces. I sit. I wait. Then I turn him – and he immediately walks politely and without hesitation out of the barn, up the drive, as far as the entrance yard of the house (my intended objective). I thank him and pat him. We walk back again.

My understanding of leadership has increased. I have had a clear, repeated, experience of the importance of *timing* in pacing and leading: even if the moment of pacing is very brief, as in allowing a moment's pause for Mouse to collect his mind and refocus after his threatening bounce before asking him to go forward, it still needed to be established. In terms of human interaction, it would be like ensuring that someone finishes stating their point of view and making sure they know you have really heard and respected it before going on to persuade them in a different way. *For a conversation, both parties have to be available to each other.*

My options for being the kind of leader I am comfortable with being have increased. That moment's pause is functional for me, too. It means I don't feel that I have to get into an argument, or win my point, or attempt to overrule (and who can readily overrule half a ton of horse…?) The pause enables me to let Mouse know that I have heard his reluctance and understand its reasons – but that I have good reasons for reasserting mine, which in the long run will work for both of us. That kind of leader, I can be. *Meaning it*, I can do, because that's about me managing myself. Only if I am comfortable within myself will my request come across as authentic to him. I have known these things intellectually, of course – but experiencing them is something else.

When I was training in NLP, I was lucky enough to be taught by Robert Dilts who, as explained previously, was one of NLP's earliest developers. He used to quote the Papua New Guinean proverb: '*Knowledge is only a rumour until it's in the muscle.*' This seems to accord with Belasik's assertion that riding is essentially a practical process – theoretical understanding is hugely valuable but it is only fully realized in practice. Now, when I am riding, I know in my muscles what it means to pace and lead.

The Gift of Influence

As a result, my partnership with Mouse has grown more subtle, more complex and more satisfying – to us both. When I say goodbye after putting his rug on he offers his face for me to put my arms around and hug.

Horses offer us many insights about intentional leadership and actual influence. If we are open to the teaching they offer us, and integrate it into our own practice, we are likely to find that they actually choose to follow where we wish to go. Not surprisingly, if we carry over these approaches into our relationships with other human beings, they too are likely to welcome our influence.

The conclusion of *Alpha Leadership* summarizes 'a new set of basic themes of leadership':

- Increase your alertness to 'weak signals'. (This means you will register important information earlier than other people.)

- Nurturing your mental agility…will help you make use of the space ahead of the curve [which] your alertness creates and should make sudden, unexpected changes in the…environment less threatening and easier to handle. (This means you will be able to take advantage of situations and react more quickly than other people do. You will also be less surprised and 'thrown' by events.)

- Successful leaders embody their goals, and align all their beliefs, values, competencies and behaviours behind their calling as…beings, rather than merely as…leaders. [I explore the concept of alignment more fully in Chapter 8.]

- We propose a tactic of 'in-course correction' – to act, then adapt, rather than to wait for the perfect circumstances before moving forward. To keep ahead of the game, take action and improve as you go.

- Less is more: do a little very well indeed.

Extrapolated from *Alpha Leadership* by Deering, Dilts and Russell, pp.206–7

The Gift of Influence

I believe that, essentially, each of these points about outstanding business leaders also characterizes the way horses naturally interact with each other – even regarding the connection between excellence and economy of effort! Observing our horses, reflecting on how they *actually* experience us and how we would *like them to*, and following these guidelines towards a new model of what it means to be a leader, can help us become more influential, with less stress, in both human and equine relationships.

The Gift of Influence

Chapter Six

The Paradox of Attention

Concentration is not staring hard at something. It is not **trying** to concentrate; it is not thinking hard about something. Concentration is fascination of mind.

Timothy Gallwey, *The Inner Game of Tennis*, pp.80–1

In his work as a tennis coach, Timothy Gallwey noticed an important paradox: concentration of the 'trying' kind can often be effortful, but when concentration arises naturally as a result of fascination it feels effortless and is often more profound, engaging someone's attention at a very deep level. In other words, what we label 'concentration' describes not one but several kinds of experience, in terms of both mental processing and mind-body states. We can focus intently in very different kinds of ways.

If concentration is – quite literally – the merging of our centre with that of whatever we are paying attention to (as in the word 'concentric', describing circles that have the same central point), what is it that attracts us and makes us pay attention in the first place? In terms of our animal history, it's probably those two powerful motivators, *opportunity* and *danger*. On the one hand, finding dinner, shelter, friends or a mate, and at the other extreme spotting something or someone that could be life-threatening. For horses, these are the 'alert' buttons that focus their minds and drive their actions even after thousands of years of domestication. This attraction-fear dilemma is one of the underlying paradoxes of their daily lives, and one that we have to factor into our own thinking if we are to communicate and work effectively with them.

Despite our human abilities to reason, self-reflect and self-manage, the twin powers of attraction and fear operate just as much on us, not least in the ambiguity of our relationships with the horses themselves. I remember as a teenager being anxious every time I rode. Saturdays meant an upset digestion, inability to eat, butterflies in my stomach, dry mouth… Just as strong, however, was the pull, the fascination, the determination – and the *love* – that impelled me out to the stables every week. Time and experience may have softened off the contrast, but it is still there. I sometimes say that one of the attractions of horses for me is that they are the only thing in life I'm scared of. What an odd statement, because I have always hated the physical sensation of heightened adrenalin in my system; yet it reflects an underlying suspicion that I am more alive when my senses are sharply in play. Having a strand of fearfulness in what I most love to do makes me feel like the Metaphysical Poets, whose work often gains its edge from the tension between their fascination with life and their equal fascination with death. Riding is where I feel *alive* – because when I'm around horses the love and the danger they represent to me invite me to be really, truly present. This is closer to Gallwey's argument (*The Inner Game of Tennis*, p.81) that: 'When there is *love* [my emphasis] present, the mind is drawn irresistibly toward the object of love. It is effortless and relaxed, not tense and purposeful.'

In this chapter I want to explore the interplay between these three concepts – attraction, danger and love – in catching and holding our attention, and to ask how understanding and working more flexibly with attention and its focusing can enrich us in our whole range of relationships, and in our lives overall. In a nutshell, I want to show how (if we let them) horses can, through the love they generate in us, help us become more aware, more flexible in establishing a richer picture of what 'reality' means, and more subtle in how we use any information available to us.

I also want to look at the different mental perspectives or templates that we have to choose between when we do pay attention. My argument is that horses can enrich our lives by challenging us to face some of the paradoxes of what paying attention actually involves. And, in accepting that challenge from them, we gain the possibility of developing greater mental agility:

The Gift of Poise

Circumstances change, but life goes on. In the business world too, leaders need to be mentally agile. They need skills in the three vital characteristics of mental agility: exploring beyond their boundaries, recognizing the limits within which creativity operates successfully, and being flexible in pursuit of goals despite encountering obstacles to their achievement.

Anne Deering, Robert Dilts and Julian Russell, Alpha Leadership, p.36

If we are to learn to be effective leaders, not only of our horses but of ourselves, we are going to need that mental agility – in spades!

What happens when we pay attention?

Paying attention actually involves several processes:

- Scanning the environment.

- Registering or noticing information that's new or that has changed.

- Attributing meaning to what you have noticed.

Only when you have done all three are you able to respond appropriately. Attributing meaning, in turn, depends on *how you understand what you have noticed*. Understanding depends on the mental template through which you manage the information. And it's here that sharp contrasts of possibilities can confront us with important choices and balancing acts:

- Do we view the situation from our own perspective (*self*) or that of the *other* being(s) involved ?

- Are we motivated primarily by the pull of what we find attractive, or by seeking to 'cover our rear' by avoiding what might be risky, unpleasant or dangerous?

- Is our focus on what's happening here-and-now, or are we taking an overview in which we relate what we're currently experiencing to the wider framework of past and future?

Scanning the environment

When I was teaching on a postgraduate certificate course for intending teachers, the student who was best at scanning his classroom had three children of his own under 3 years old. For him, scanning a group of more than thirty 11-year-olds in a busy classroom was no more difficult than monitoring the action in his own kitchen or living room. So the classroom situation never got away from him, as it did from fellow students who did not have his domestic 'training'.

Scanning is a process of appraisal that is without inherent urgency, and many animals are excellent at it. If you watch your cat or dog, they may appear to be dozing, yet often their ears are swivelling in response to sounds and, if they need to, they can go from sleep to full alertness in fractions of a second. Horses are just the same. Scanning helps the predator to see, hear or scent possible prey; and it helps prey animals to mobilize in good time to run away.

How does scanning work physically? There seem to be two key features. The first is that attention is not deliberately directed but flexible and responsive in its nature. This is the 'not staring hard' that Timothy Gallwey is talking about: he found in his work as a tennis coach that *intention* often gets in the way of *performance* (it tends to 'fix' the mind and the visual focusing), and that players improved most when they learnt to *notice* in a more inquiring, receptive, less deliberate way than they had been used to. The second point is that the scanner's eyes aren't specifically focused but instead remain defocused or in 'soft focus'. This means that they are better able to notice – and *then* to focus on – anything that changes – for it is *change* that signals the approach of something new, whether it's danger or dinner.

Horses, of course, have wide peripheral vision built in. The placing of their eyes gives them coverage of most of the area around them: their only blind spots are a small triangular segment of space directly in front of their eyes and a larger one behind their tails. These blind spots can normally be compensated for by the flexibility and rapid movement of their necks – and in addition, of course, by the complementary alertness of fellow herd members.

In riding and training horses, we need to cultivate a sympathetic awareness of the instant alertness and ready anxiety that their nature

The Gift of Poise

Katie is teaching her foal by example: from their half-closed eyes it seems that nothing visual has caught their attention, yet their ears are pricked and their heads angled in the same direction, so they are still monitoring their environment.

makes them liable to. But, like my student, we can also learn to use scanning for our own benefit. To begin with, it's valuable to become more aware of how *we* monitor our surroundings and the effect it can have on us. (When I was working as a therapist I was often asked to help people who had panic attacks and I learnt that, whatever their personal circumstances, such people very often had especially wide peripheral vision. You can discover how wide your peripheral vision is by getting a friend to stand behind you, first outstretching their arms, and then slowly bringing them forwards around you in an arc – you look straight ahead and sing out as soon as you see movement out of the corner of your eye. Some people will notice their friend's moving hands even before the hands are in line with their shoulders; others may notice nothing until they are beyond the shoulders and virtually in front of them. Wide peripheral vision admits more information, so sometimes it may be sheer visual overload that tips the person over into having their panic attacks. Standing by the indicator board at a major international station can have the same effect, because there is simply too much infor-mation outside of your intentional focus for you to process! Your unease or downright anxiety is the result of visual overload.)

The Gift of Poise

On the plus side, peripheral vision, rather than straight-ahead vision, is very useful to you if you want to become more sensitive to changes in your environment. Allowing yourself to get stuck in tunnel vision – literally and metaphorically – means you miss so much! If you want to observe how someone reacts in a difficult situation, you will literally notice more if you watch them indirectly: you will catch not just their obvious changes of posture and expression, but also small and subtle changes in their muscle-tone or colour. Watching your horse in the same indirect, unfocused way will help you get a much finer appreciation of how he is feeling and how he may be responding to your requests, commands and teaching. There's a further benefit, too: when our eyes are defocused we almost automatically enter an altered state of consciousness, one in which we are aware without anxiety and alert without urgency. All our physical senses are potentially online, as are both conscious and unconscious mental processing. We are in a state of much greater *availability*.

Shane's eyes are defocused, but it's obvious from the connection between her and her horse that they are in rapport with each other. In such a state, she is poised to notice the slightest change in him, in herself or in the environment, without tension yet able to respond at once if needed.

The Gift of Poise

Registering or noticing information that's new or that has changed

When we allow, or actively encourage, ourselves to become available like this, we gain some important advantages.

1. We become alert to what have been called 'weak signals'.

   Anticipation starts with detecting weak signals. Everyone can hear a shout, but only those with exceptional sensory systems can hear the barely audible whispers where most of the opportunities and timely warnings lie.

   Anne Deering, Robert Dilts and Julian Russell, Alpha Leadership, p.15

   You do not have to take on the role of predator or prey to learn to tune in to such signals: whatever your role, extra sensitivity gives you *more time and more choice*. Experienced riders often say how slowly time seems to go when they are competing: novice riders rarely feel like that! It has also sometimes been said of exceptionally talented jockeys that they never seem to be short of room – even in large fields, at racing pace, they avoid getting shut in. In part, this is a result of their differing abilities to receive and respond to weak signals. This was what helped my student in his classroom. It's what enables good 'people people' to spot what others are thinking and feeling almost before they know it themselves, and thus respond rapidly as well as appropriately.

2. If you have spotted the early signals of change in the environment, you will be quick to notice when the same signals are repeated and start to indicate a pattern. The slight sound in the distance that makes your horse prick his ears and lose concentration for a moment may be the first indication that the hunt is coming your way, that a huge combine harvester is coming across the fields, or that a helicopter is a mile or two off and moving in your direction. If you have trained yourself to notice slight changes in his attentiveness and muscle-tone you will be cued in to receive further information that confirms or dismisses what you first noticed, and to seek explanations in good time to prepare your own reactions. You will be all the better prepared

At first glance, it seems that Gaz is paying more attention to the judge than to his young stallion; but his hidden left arm and the whole of his left side are in contact with the youngster, part of a kinesthetic conversation that says, 'I'm still with you.'

for a timely escape, for relaxing again, or alternatively for redirecting your horse's attention in a more inviting (even demanding!) way. In other contexts, if you have learnt to notice tiny changes in other people's facial expressions, voice-tone or language patterns, you are more likely to spot opportunities for steering or remedying interpersonal situations before they go crucial on you. If you are in business, you stand a better chance of sensing changes in the marketplace ahead of others and of keeping the edge on your competitors – once you have decided how to respond, that is.

Attributing meaning to what you have noticed

Scanning and noticing are the foundations for attributing meaning and for taking action. The horse who shies at the hosepipe that wasn't there yesterday has noticed today's difference and reached the conclusion that the hosepipe is, or might be, dangerous. When my husband Leo and I

The Gift of Poise

were hacking home along an unfamiliar road some years ago on the two horses we had at that time (one older and one just an inexperienced youngster), we came upon a turn in the road which had **SLOW** painted in huge white letters across it. Our old horse, Lolly, looked down briefly and plodded on: our youngster, Vals, hesitated, then attempted to jump over the whole band of letters. Lolly had road markings stored in his memory banks; Vals, who was new and green from Russia, almost certainly didn't. The same stimulus meant different things to each of them, because they were processing the information in relation to different experiences and expectations.

This is an important lesson we can transfer from horses into our everyday lives. It is your *interpretation* of the information that tells you whether the stimulus means dinner or danger – and the *accuracy* of your interpretation depends not just upon your knowledge and experience but upon your *mindset*. For it is your mindset that leads you to observe selectively in the first place, or to interpret selectively in the second.

Mark Rashid gives some fascinating examples of mindset and its effect. One horse-owner on his clinics thought his horse was 'resisting' when, from the horse's point of view, it was more likely that he was impatient at being asked to repeat a manoeuvre he already knew, with no apparent purpose other than that, in his owner's mind, repeating it would demonstrate 'obedience'. Many other owners tried to get their horses to learn a new movement (e.g. backing up) but because they were mentally fixated on getting a clear, complete response they failed to see the slight signals that showed the horse had understood and was willing to attempt the manoeuvre. Rashid calls this tiny indication of willingness *'the try'*, and points out how reinforcing it can be for both horse and rider when the rider does spot it – and how demoralizing and confusing when the rider doesn't. Failing to 'get' this is one of the biggest reasons for problems in schooling.

Alpha Leadership draws our attention to a problem inherent in weak signals such as 'the try':

An organization's sensing system will only give it a real edge if it can detect signals before they become clear and unambiguous. This is a problem… Because, like a whisper, a weak signal may be ambiguous. There may be other signals that contradict it, its weakness may make it hard to understand, or may simply mean it signifies nothing. It may be tantalizingly equivocal, but it must not be ignored because, if you wait until things become clear before you act, you will act too late.

Anne Deering, Robert Dilts and Julian Russell, Alpha Leadership, p.19

Our relationships with horses offer us plenty of examples of how we can wrongly interpret – and therefore wrongly respond to – their weak signals, not only because of the signals' inherent ambiguity but primarily because we receive them with a mindset that limits the possibilities of both reception and interpretation. Equally, though, it shows us how acting quickly can save our bacon – or even our lives.

Whatever the potential risks of wrong mindset, without *some kind* of mindset or sorting mechanism, we cannot make sense of the sensory information we receive. We can't spot a 'try' unless we first have the concept in our minds and are looking for it. And one of the paradoxes of attention is that our major sorting mechanisms give us not clear-cut but seemingly diametrically opposed choices.

In making sense of all incoming information, we have to sort it somehow, and NLP has identified a number of governing mental processes (sorting mechanisms) that help us do just that. In terms of understanding attention, three of these are particularly relevant; we could see each as being rather like the sliding scale of a dimmer switch – the scale has two ends, each of which represents an extreme position, and we will probably end up somewhere in between. These mechanisms are:

- Understanding viewpoint – the continuum between *self* and *other*.

- Understanding motivation – the continuum between *away-from* and *towards*.

- Understanding time-frames – the continuum between *in-the-moment* and *overview*.

Understanding viewpoint – the continuum between self and other

Let's begin with some common problems people face in their relationship with their horse. One of the commonest mindsets afflicting riders is the anthropomorphic attitude – the assumption that horses think and

The Gift of Poise

behave like people. From this standpoint, difficulties and misunder-standings are seen as 'resistances' or 'problems' or 'evasions', when in fact they may simply demonstrate that the horse hasn't understood, or is showing that – as is frequently the case with 'laziness' – he is simply being a horse. In his book *The Nature of Horses*, Stephen Budiansky reports from research on the mechanics of equine movement that:

> At each gait there is an optimal speed, which horses naturally tend to favor, at which energy consumption is minimized. When traveling at these optimal speeds, the energy required to move a given distance is the same whether at the walk, trot or gallop.
>
> Stephen Budiansky *The Nature of Horses*, p.202

In being economical with his energy, the domesticated horse is actually responding to an inbuilt programme to conserve energy which presum-ably goes back to a genetic imperative to have some reserve in case of the need to take flight.

In the previous chapter I explored the issue of influence. Riders and trainers often seem to feel that the only choice they have is between hav-ing an 'obedient' or 'disobedient' horse. The argument of this mindset goes something like this: *'If he does what I ask he is submissive and I've succeeded: if he doesn't he is disobedient and I have failed.'*

Underlying this rather dismal dichotomy is the first of our important sorting mechanisms, one that matches what happens against a somewhat different, and deeper, set of choices. When you assess the response you get, whose perspective are you taking – your own perspective (*self*) or his (*the other*)? When you have a row with someone at home or a colleague at work, are you justifying your perspective largely because it's yours, or are you wondering how the argument might have seemed to them, or how it might have affected them? As their boss, are you assuming you need to 'tell' or 'direct' them in their work? As their sub-ordinate, are you assuming they won't be receptive and will think your perceptions, your ideas and your initiatives are wrong, stupid, or chal-lenging *just because you are in a subordinate position?*

If you are to have a more complex and therefore richer understanding of what's going on, you need to be able to imagine how things are from the viewpoint of *the other* as well as the more natural viewpoint of *self*.

The Gift of Poise

Whether you are interpreting a situation, spotting weak signals, identifying patterns, or deciding how to take things forward, you will need
to find a temporary perching place somewhere on that *self-other* continuum. Better still, you will be able to gather the fullest range of
available information before making any decision or taking any action,
by sliding flexibly along the continuum just as if it were the slide of that
dimmer switch: 'reality' is more complex than simply opting for either
extreme.

Understanding motivation – the continuum between away-from and towards

When we think about motivation, we usually ask what *drives* someone
or *what they want*. The words themselves tell us that people can be
motivated negatively or positively: they can seek to avoid, or get away
from, something or someone they dislike, or they can seek out something or someone they actively like or aspire to. (The danger-dinner
interpretation of change in the environment is an example of this.)
Some of the thinking about horses' motivation seems to fit neatly into
this, because it argues that horses' fearful nature and flight behaviour are
those natural to prey animals (away-from motivation), and are inher-
ently contrasted with the dominating (towards) instinct that human
beings have as predators.

However, this is too simplistic a view: recent writers on equine society
have pointed out that humans haven't actually preyed on horses for
thousands of years, and that most equine behaviour supports the view
that, if anything, they see humans as allies or even as honorary herd
members. Marthe Kiley-Worthington and Mark Rashid have both
argued from their observations of natural herd behaviour that, amongst
themselves, horses actively seek out friendship, support each other, and
prefer co-operation to conflict – all motivations on the *towards* end of
the continuum.

The Gift of Poise

Understanding time-frames – the continuum between in-the-moment *and* overview

It's obvious that when you're riding you need to be paying attention to what's happening in the present; but every moment requires you to make some kind of judgement as well as some kind of response. Is your horse frightened when he spooks, or is he napping because he'd rather be going home? Mouse's reluctance to go past the barn and down to the school on one occasion (see previous chapter) probably originated from his learnt understanding that after a hack you return to your box. The same behaviour in the same place another day probably resulted from my asking him to go down to the school *in front of* Gordon, whom he usually *follows* because Gordon is his alpha horse. Neither instance was an example of naughty nappiness, though the issue and the frustration felt much the same as if it were. If you can find a meaning in your horse's terms, you are halfway to looking at your experience in another light: your assumptions, your feelings, and probably your subsequent actions, have all been 'reframed'.

As these examples show, we cannot restrict ourselves just to the present. Being in the moment is exciting, enriching, frustrating, depressing, exhilarating…*in relation to* the meaning you make of it – and meaning always includes the past and will probably also include the future. It's a bit like the zoom lens on your camera. Before you zoom, you get an overview of the scene as a whole. If you zoom in, you get the here-and-now detail. The detail is meaningful as a choice from the bigger picture. The bigger picture gives you a way of understanding and evaluating the detail. It's the same with the tension between experiencing this moment here-and-now and experiencing it in relation to the past and future.

Managing issues that arise between you and others, whether those others are horses, families or work colleagues, will require you to set immediate behaviour in context (the long view) and also to use the long view to help you understand why they, and you, feel as you do here-and-now, and work out how you can act in order to make progress towards your preferences and your goals.

Wise horsemen have pointed out that 'in riding there is no neutrality'. By this they mean that the horse learns *something* by every response you make. If you praise him, he learns he's done well; if you ignore him, he learns that what he did was acceptable; if you chastise him he learns he did wrong; if you repeat the task with the same instructions he may learn to understand them through repetition; if you repeat the task with clearer instructions, or break it down into smaller chunks, he learns that although he somehow got it wrong the first time he can get it right another time. People are actually much the same… In managing yourself and others, and yourself in relation to others, your ability to zoom with flexibility and finesse is a key secret for success.

Attention and love

But there's more: the magical element in creating and maintaining attention – Gallwey's concept of *love*. Love is more powerful and more enabling than any specific choice: it builds bridges between self and other and modifies the stark motivational contrast between away-from and towards. It links past, present and future. Love, fascination, amazement, wonder… This is what makes us bother to relate to other beings, whether human or animal. Love is what makes entomologists study moths, palaeontologists study rocks, musicians struggle with strings and chords and harmonics and pitch, as well as what makes us seek out partners, create families and venerate ancestors. It's why we hate conflict. It's why we keep trying. It's why we have horses in our lives.

   There's something very pure about this fascination, this loving. It's about being at the edge of our experience, the skin of our very selves – *because that's both where we are exposed and where we can be connected.* When I think of attention, or concentration in this sense of overlaying the very centre of myself on something/someone that really matters to me, I have a sense of spaciousness that's like the kind of learning Timothy Gallwey describes. It's not easy – far from it. In my work as a coach I sometimes feel that what I think of as 'I' has disappeared: in being as vitally present as I can be, I am distilled down to my eyes and my ears. When I'm out hacking in harmony with my horse, fully attentive and fully open to all my senses and to the world around, or when we're working in the school and for a time I have that sense of moving so closely with him that there's no mental or physical space between us, I have lost my sense of separateness and found a sense I can only describe as one of *poise*. Poise is the moment of readiness where all one's faculties are effortlessly accessible. This is the place from which we can, in Gallwey's words: 'Close the unnecessarily wide gap between our potential and our actual performance and…open the way to higher and more consistent levels of enjoyment.' (*The Inner Game of Golf*, p.26) And it's also the place where we can: 'Understand [the other] well enough to be able to help him achieve his goals while he helps you to achieve yours.' (Anne Deering, Robert Dilts and Julian Russell, *Alpha Leadership*, p.113)

Poise is a moment-by-moment balancing act, yet it gives us a secure platform from which we can begin to overlay our meanings, needs and aspirations with those of the other – whether person or animal – and

Hacking around the farm on a warm summer evening, Lolly and I are so much attuned to each other that we are also connected to our surroundings in a special and much wider way. For me, this was a magic experience in which Lolly connected me with the full world of my senses and of the countryside. Leo has managed to capture both the detail and the sense of wider horizons that were special about this experience.

The Gift of Poise

find a fit that works for both. In Rashid's words, it allows us to begin to 'blend' with one another. In *Life Lessons from a Ranch Horse* Rashid talks about the difficulty he had in forming a calm partnership with his horse, Smokey – *until* he realized that Smokey's natural restlessness needed an outlet in problem-solving and devising new tricks. And he describes how a conflict of beliefs between himself and a horseman at one of his clinics threatened to disturb and anger both of them – *until* Rashid: '…found a way to get along with him without compromising my beliefs… We both accomplished our goals, and we left without any hard feelings toward each other. In the end, maybe that's what blending is all about.' (*Life Lessons from a Ranch Horse*, p. 103.)

When we allow horses to help us focus our attention in this way, it's like catching the sun's light through a lens: the light of our concentration becomes brighter – bright enough to make lasting differences on how and who we are.

> About two years after I met Buck… I noticed a gradual shift in the way I was handling myself… An overall slowing down of just about everything I did. Oddly enough, the slower I went, the faster things seemed to get done. The less pressure I applied, the quicker the response I hoped for. The fewer words I used, the more people listened. The less I talked, the more I heard. The less I worried, the more I could focus. The more I could focus, the calmer I stayed and the easier it was to make informed, positive decisions about things.
>
> Mark Rashid, *Life Lessons from a Ranch Horse*, pp.76–7

The Gift of Poise

Chapter Seven

The Challenge of Humanity

Riding could be compared to playing a game of chess with one-self, either winning or losing at one's own hand. But the marked difference between the two is that in riding the horses are on our side helping us make the right choices – if we avail ourselves of their counsel, that is.

Erik Herbermann, *A Horseman's Notes*, p.12

When I asked my husband Leo what horses did for him, his reply, quoted earlier, sent shivers up my spine: '*They challenge my humanity*.' The very idea was so profound and so disturbing. What did he mean? And why did I immediately sense that if horses did this for him they probably did the same for the rest of us? In this chapter I'm going to tease out what it can mean for us when horses 'challenge our humanity'. Leo's story comes first, both as a way in and as an example.

The challenge for Leo

Leo began riding when we were given our first horse, nearly eighteen years ago. He was 52 and had never ridden. Typically, he thought that if we were going to have a horse he would learn to ride – so he did. Leo has excellent balance and is a conscientious and effective learner, so he really applied himself. He doesn't have a natural sense of timing, and in NLP terms he relies on vision and hearing rather than kinesthetic sensing or 'feel', both of which have made it harder for him at times. But he really cares about riding properly; he loves the horses and has virtually no fear around them or about doing anything with them. So he creates an atmosphere of relaxed calm and is willing to have a go at most things (he even rode side-saddle once).

Nonetheless, he can feel really challenged – more so since our old horse, Lolly, died. Lolly had been with us for thirteen years and was very responsive and steadfast, as well as intelligent. Though he could be quite sharp, he was very rider-focused and each of us developed a trusting and confident relationship with him. In his later years, one of the delights we shared was hacking round the fields, interspersing quiet walking with bursts of lively canter that lasted for as long – or as little – as Lolly wanted to continue. The next horse we had, Bertie, seemed promising when we bought him, but though he was pleasant to ride in the school he was aggressive in the stable and became nappy out hacking. Leo managed Bertie in the stable with an effective mixture of firmness and persistence: he didn't get angry and he wasn't scared. But he found the nappiness frustrating as well as disappointing, and not being able to get Bertie's co-operation made him feel that he was inadequate. After about a year we agreed that Bertie was not the right horse for us, and that was when we got Mouse. As I've already described, Mouse can also be prob-lematical out hacking and, though his reasons are quite different from Bertie's (and his character is benign and co-operative, not self-contained and hostile), at the time it often *feels* much the same when he refuses to go forward: frustrating, disappointing, and as though we are not adequate as riders to deal with the challenge he's presenting.

This is the challenge as Leo experiences it:

…to get what I want from the horse by disciplining myself. There are always people around who spend all their time on this, as opposed to the three hours a week that I spend on horseback as an amateur rider. I can't draw on my experience of good human communication, for example, modelling the way my parents managed us, because horses speak a very different language and have an embedded evolution which is very different from ours. In my working career I was puzzled when colleagues wanted me to 'bang people's heads together', because my way through the human world is through negotiation; but even so, how do you negotiate with a horse? You cannot rely on having shared goals with

The Gift of Self-acceptance

animals – the horse has completely different ones. You don't clip a child round the ear and hope that will teach him something, but you might have to be physical with a horse to help him understand what is wanted. At its most frustrating, I feel I want to be humane but it's not working. I want Mouse to trust me out hacking, but he doesn't yet.

Leo's difficulty is that of all humane, caring riders: they want to be friends with their horse, but sometimes their partner's very 'horseness' gets in the way of this. The potential difficulty here is that a horse's behaviour, in some circumstances, can seem like a real let-down or 'betrayal of friendship'. My editor, Martin, commented: 'I once described it to someone as being like seeing someone you know in the pub, walking up and saying, "Hello mate, can I buy you a beer?", and them turning round and punching you in the face.' The feelings that the horse's behaviour immediately stirs up are likely to be as powerful, and the strategies for dealing with the situation only as effective, as the earliest time you felt like that: for most of us, the shock and the obscure sense of betrayal that Martin describes take us back to playground scuffles and disappointments. We are effectively disempowered. As a result, the difficult challenge in such circumstances is not to take offence and react angrily to the horse, or alternatively to give up the immediate goal and feel inadequate.

For Leo, the challenge of managing Mouse echoes childhood experiences of trying to cajole his five younger siblings to do things when he was himself quite young, and also of the frustration and puzzlement he felt many years later as a manager at work when negotiation wasn't effective and he couldn't bring himself to influence his colleagues through more forceful means instead. As I shall explain more fully in the next chapter, trying to act against your own beliefs or character makes you even less effective – and miserable into the bargain: friends and trainers who told Leo he 'ought to be firmer', or that Bertie (or Mouse) needed to be 'made to' do this or that only added to Leo's difficulties because their solutions involved thinking and acting like *them* not like himself, and so compounded his sense of inadequacy. Leo has to find solutions that feel right to him as well as convincing Mouse. (As, of course, do I.)

This dilemma is one which may begin with us and our horse, but rapidly brings us up against *ourselves*.

The Gift of Self-acceptance

The interesting thing about this photo is the contrast between Leo's concentrated, 'trying hard' expression and Mouse's obedient, correct attention. Mouse seems thoughtful but Leo seems to be experiencing a sense of challenge which is more emotional than physical.

The strands of the challenge

It's important to step outside the (quite natural) praise-blame framework here. We are not dealing with failure or success, even in the strictly limited terms of things turning out as we'd like. The success/failure framework is very limiting even when we judge that we have succeeded – because it doesn't make us any more resourceful: feeling pleased with yourself stops you from exploring further options or asking deeper questions, and in the long run these are what make you more versatile and more effective. The challenge, as Herbermann describes it, is a challenge of strategies and moves that takes place within each of us. As Leo said, 'everyone will have different conflicts with horses' because horses challenge each of us according to our individual natures as well as theirs. Examples of this are:

- **The challenge of communication**. We need their co-operation and have to find ways to explain what we want in terms that are meaningful to them, not just to us. This challenge involves us in developing greater sensitivity and control over our own bodies so that we can aid effectively and consistently, and it also involves us in clearing our own minds of emotional clutter and 'white noise', which can distract or confuse our horses and block the channels of our communication with them.

- **The challenge of understanding and adapting**. We can't expect that horses will share our goals – so we have to find ways to make what we want pleasant or worthwhile to them. This involves us in stepping imaginatively into their shoes (taking second position) and trying to create experiences that will work for them, rather than simply attempting to impose our own will.

- **The challenge of echoing or repeating our old history**. Often, interaction with horses brings up issues we have faced before in our human relationships – bringing up similar feelings and judgements about our capability and adequacy. Management trainers and therapists who use horses as partner-assistants in their work encounter these 're-runs' all the time and, in fact, welcome the opportunity to help their human clients find new strategies in their thinking as well as their actions.

- **The challenge between differing parts of ourselves**. We can experience problems with our horses as issues, not just between them and us, but also, at a deep level, as issues between us and ourselves: external conflicts become internal conflicts. To refer back to an example given in Chapter 2, I have worked with many riders whose horses can be tense or spooky in competition – but often when we unravel the situation this is actually the *result* not the *cause* of the rider's own feelings, beliefs or behaviour. Riders' own anxieties or sense of perfectionism about competing can cause them to distance themselves emotionally from their horse and 'grab him' physically as soon as they enter the arena: it's not surprising that he then loses confidence and becomes tense.

The Gift of Self-acceptance

How horses challenge our deepest selves

How do horses get us in so deep? Horses are drawn into relationships with us, just as they are drawn into relationships with each other. A horse has his place in the social order of his herd or field-group: this lends him the advantage of their eyes and ears and he benefits from their special skills and protection. He finds friends for mutual grooming and head-to-tail fly-swishing. It seems likely that such relationships, from a horse's perspective, elicit and reinforce his individual identity as he contributes in his own way to the everyday functioning, safety and continued social life of his group. He knows who he is and where he fits. He knows the same about each of the other herd members, for we know that horses recognize each other as distinct individuals, and that in their herd, field or stable groups they build networks of overlapping loyalties and hierarchies and engage in quite complex negotiations for space, power and affection.

We can't know, of course, just how horses view us, but it would seem that there are enough similarities between humans and horses for them to find it quite natural to include us in some way in their social circle. Certainly, there are obvious benefits from contact with us! It could even be that female horse-people are like the alpha mares of the wild herd, in that we protect, nurture, train and discipline our horses. Since horses don't depend on words to understand or influence each other, they quite effortlessly understand our raw, genuine communication even when – perhaps especially when – it differs from what we like to think we are 'saying'.

So horses offer us both a personal quest and a promise. The quest means that, in their presence, we have the option of discovering ourselves: the promise is that if we accept this encounter we find more than we feared, and often more than we dared to hope. This is what I think Temple Grandin is implying when she says: '*All animals make us human*' (*Animals in Translation*, p.306) and why a friend, replying to my question: 'What do horses do for you?' wrote: '*They give me…an empathy with myself*'.

Perhaps this is how the theatre saying: '*Never work with children or animals*' arose: both groups can embarrass us profoundly because they respond to us with directness and without disguise. Essentially, of

Every time I get on
a horse I am aware
of the promise and
the quest that the
encounter offers.
Adjusting the stirrups
before mounting a
relatively unfamiliar
horse was one of
these moments of
possibility...

course, *we are embarrassing ourselves* – it's just that, being how *they* are,
they show us up for what we are. Our attempts at manipulation backfire;
our short-cuts turn out to be winding diversions; our half-truths are
more than half lies. *They can tell.*

There can be a safety and a relief in being seen through – so long as
we don't feel judged. Often, the image we see in the mirror held up to us
like this is that of our own divided self. Equally, it can be that self at
peace, at one, comfortable in its own skin, because there is no joy so
profound as being recognized – *and approved* – for what one truly is.

This is the arena in which equine-assisted therapies have begun to
work: people with physical, emotional and social difficulties have all
found help and healing through supported interaction involving horses.
But the challenge and the opportunity to recognize, accept and find new
ways around limiting physical, emotional and interpersonal functioning
that horses offer human beings are all available to anyone who comes
into contact with them – *provided that the person is willing to engage with*
themselves.

The Gift of Self-acceptance

Before attending the equine therapy course, this lady said that she was fearful of horses, and perhaps of other things in her life. At this moment she is returning from the long journey to the other end of the arena. Still a little tentative, she is radiant with the realization of what she has been able to achieve.

Actually, the choice is loaded in favour of engagement, because it's hard to dodge what a horse confronts you with. Humans can all too easily wriggle out of dealing with other humans. Quite frequently, in my work as a coach, I find myself being asked to deal with people whom their managers find 'difficult', 'high maintenance', 'poorly motivated' or any one of a number of similar adjectives. Sometimes I'm told that one professional has asked for another to be left out of a project that would involve them working together, 'because he really irritates me', 'because she has an issue about power' – or information – or delegation – or

The Gift of Self-acceptance

whatever. By labelling the difficulty as 'belonging' to the other person, it's possible for the complainant to dodge the possibility that he or she might also be putting in their own two-penn'orth. Instead of looking at *your* difficulties, *your* limitations, just pass the problem on to someone else's manager, or perhaps even get them a coach…

Such responses are not so easy where horses are concerned – though some owners do choose to send their 'problem horses' away for expert 'coaching' and re-schooling, only to be surprised when the new behaviours so carefully schooled in are not always maintained on the horse's return. You can tell yourself that the issue, whatever it is, is 'his stuff', not 'your stuff'; but that leaves you little better off in terms of what to do next. In fact, it can actually sharpen your sense of the dilemma you are facing.

The fundamental challenge faced by every rider is to find a way of building a relationship between their own uniqueness and that of the horse. It's that bond that shines out in this photo of Leo and our young horse, Hawkeye.

It's an uncomfortable choice, whether the relationship is with our children, our colleagues or our horses. We feel the discomfort, actually, because the dissonance, the uneasy feeling that both options are unpalatable, is within us: it's *intra-personal* not interpersonal.

In a nutshell, horses challenge our humanity in three main arenas: physical, emotional and mental.

The Gift of Self-acceptance

Physical challenges

Learning to stay on, steer, stop and go more or less reliably is only the first of the physical challenges horses offer you. The American Equestrian website, outlining the nature and befits of *hippotherapy* (physical therapy involving horses), reported that Dr Daniel Bluestone, a paediatric neurologist at the University of California San Francisco, in research using MRI scans over time, found that 'the repetitive movement of riding prompts physical changes in the brain'. In a Discovery Channel documentary he said:

> We think that hippotherapy is effective in helping repair networks within the cerebellum and within the motor system up in the cerebellum. The pathways within the brain that facilitate a particular movement become reinforced over time. The more pathways you reinforce, the better the brain compensates and the better motor function can improve.'

www.americanequestrian.com/Reference/ref.hippotherapy.htm

Boiled down, this suggests that *riding improves physical function because its repetitive movement beneficially affects brain structure and function.* If riding has these effects on people with physical and neurological problems, it makes sense to assume that it also has similar benefits for those of us fortunate enough to be without such difficulties. Think of how many movements you might be making every time you sit on a horse: you move in three planes (forward-back, upwards-downwards and side-to-side) in relation to three frequently used yet quite distinct gaits (walk, trot, canter) and perhaps the occasional gallop, varied according to the physical structure, muscle-tone and attitude of the individual horse carrying you. At the same time, you need to centre your own balance with that of a powerful creature, and to maintain that balance over time and through changes of speed, direction and terrain.

So riding is a substantial physical challenge – and a physical education. Once we got into dressage as a family I remember marvelling at how much I now noticed, and how much I had learnt, about my own body. The neat phrase 'achieving independent hands, legs and seat' represents, without in any sense explaining, just how much physical awareness and self-management are involved. Our local physiotherapist was quite clear about the benefits of riding, despite his experience of the accidents and injuries that riders suffer: 'It makes people fit in their bodies and fit in their minds', he said.

If we are coached supportively, we learn to understand and work with our physical quirks. We try to assess the inequalities caused by right- or left-handedness and to become more ambidextrously able. We notice how we find it easier/harder to bend one way than the other, how we sit more heavily on one seat-bone than another (surely not all saddles can be harder, or all knicker elastic more roughly sewn in on that side...?), how one shoulder dips more than the other, one hand is more sympathetic and connected, while the other is more grabbing and less giving. We notice these things initially because of how horses respond to us, and then we internalize that information and grow in our ability to monitor how and where we are (a skill, known as proprioception, that is to an extent naturally inbuilt, but which riding substantially enhances).

The beauty of taking on this challenge is that, the more we work with it, the more able and subtle our physical mastery gets. Something as slight as an outward breath from the rider can indeed signal and then bring about a downward transition from the horse which involves the redistribution of half a ton of muscle and mass... Becoming aware of your position in the saddle can improve your posture in the office, in your easy chair, at the dining table, in your car...

Emotional challenges

Horses challenge us emotionally in their own right, because they engage us as beings to love, care for and train. They can evoke anger, anxiety, distress, frustration, joy, elation, irritation, and calm well-being, as the paired phrases on the endpapers express so eloquently. Over and beyond this, though, when we are around them we can feel angry with ourselves, anxious about ourselves, distressed for ourselves, frustrated with ourselves, joyful in ourselves, elated about ourselves, irritated with ourselves...and also feel from precious time to time that all is right with ourselves.

This is the minefield that equine-assisted therapy helps people navigate.

Ariana Strozzi describes how working with horses showed one client how she was undermining herself in the wider arena of her life:

The Gift of Self-acceptance

She learnt that by denying her success in the arena, she was also denying Sadie [the horse's] success. Her interest in the horse feeling successful helped her pave a new path towards self-acceptance. Her capacity to care for another was the very ground from which she could learn to care for herself. She saw that by negating herself and not fighting for her own dignity, she was consequently letting others in her life down. The horse was a visual example of those 'others' in her life. She then realized that the way to take care of their dignity was to acknowledge her own strengths and successes.

Ariana Strozzi, *Horse Sense for the Leader Within*, p.105

As we interact with horses, we cannot but draw upon our past experience of interacting with other beings in our lives, repeating some of the same patterns and strategies. Yet familiar old patterns can be ineffective for two reasons: first, because horses think, react and behave differently from people because they have evolved differently; second, because the strategies each of us uses will have been skewed according to our own life-history. As we attempt to interact with horses, we are brought up against the limitations of our own experiences.

I've already described in Chapter 5 how horses show me the limitation of one of my strongest skills: while I am very good at building rapport with people and 'getting inside their shoes', taking the same length of time to pace a horse as a human can confuse the horse rather than help or encourage him. Of course rapport is needed with horses, but whereas people who are sufficiently in rapport will, with encouragement and in their own good time, begin to lead themselves, horses will not and do not. Therefore, my ability to pace can be a liability when I am with horses. (It isn't invariably an advantage with people either: sometimes it can come across as diffidence, passivity or a willingness to let others take the lead.)

Horses also confront me with another of my limitations: the fear of physical danger. I am not a bold rider; I am not an explorer. I do take risks in life – but they are usually emotional or intellectual. Riding makes me confront the physical side of myself three times a week! It makes me compassionate towards my fearfulness (and thus towards that of other people) while at the same time constantly pushing my boundaries. Allowing myself to avoid danger (for example, by doing only

The Gift of Self-acceptance

flatwork and doing it in the protected environment of the school) would also cause me to shrivel. Talking myself through situations that would seem quite risk-free to many other riders is one way I can keep myself alive, alert and at the margin of myself – that edge of experience, knowledge and skill where each of us grows the most.

Mental challenges

> Our powers of the mind are no longer just used to gather and process information from the world around us, but have evolved into one of our greatest weapons against ourselves and others. Connecting with the mind of the horse, whose instincts are intact, lets us regain our own sanity and peace of mind. Observing the instinctual powers of the horse at work reminds us we are capable of reclaiming ourselves, if we can revive our dormant instincts and awaken our intuition.
>
> A.V.R. and M.D McCormick, *Horse Sense and the Human Heart*, p.119

While it can be naïve to assume that 'simpler means better', I think that these writers are usefully reminding us that we regularly need to touch base with ourselves at an instinctual level as a way of counterbalancing and testing the value of our more sophisticated interpretations and judgements. As he spoke to me about his sense of being challenged, Leo said: 'People round the horses confront you with different explanations and different goals and "oughts", as well as different strategies.' You can get confused and even immobilized by too much theorizing: sometimes you need to ask what your own gut response is. It will not necessarily be better than what intellect (your own or others') has to offer – but it is a necessary part of the picture – *because it is part of you*. Temple Grandin reminds us that our brain has evolved in three distinct layers: the earlier two layers we have in common with animals – only the most recent layer, the cerebral cortex, is distinctly human and distinctly rational. It can allow us to manage our instinctual responses – *and it can also cut us off from understanding them*, let alone living effectively with them. Animals present us with a simpler but nonetheless authentic set of possibilities, some of them close enough to our own to remind us of parts of ourselves we can and should claim and value. Fear, apprehension, liking, loyalty, comfort, trust might be some of the values we can find again in this instinctual side of ourselves.

> When you are stuck, you're not likely to make progress by using competence as your tool. Instead, progress requires commitment to two things. First, you need to dedicate yourself to understanding yourself better – in the philosophical sense of understanding what it means to exist as a human being in the world. Second, you need to change your habits of thought: how you think, what you value, how you work, how you connect with people, how you learn, what you expect from life, and how you manage frustration.

> Peter Kastenbaum, quoted in Ariana Strozzi, *Horse Sense for the Leader Within*, p.47

It is perhaps because of this that, in many sports, some people excel even though they lack the highest degree of technical competence, while on the other hand some great technicians have had limited success. Sheer technical competence is not in itself enough.

If you are only able to talk one of the many 'languages' you have inherited – the language of words and rational analysis – you are likely to be puzzled, frustrated, even at times depressed by the barrage of information coming to you from both outside and inside yourself that is not coded in this straightforwardly intelligible way. By their very inability to speak this language, horses force us to take notice of the other languages of our lives and to learn to communicate better using the full range of our capability. As you tune in to messages coming across these other airwaves, you enter a much more complex world of sensation, enriching your experience even as you recognize its additional complexities. By accepting the challenge horses offer you at this level, you cannot but wonder about the bigger questions. The short-focus question: '*Why can't I get him to do this?*' opens up more daunting – and more exciting – questions about what life – and your life in particular – means to you; about purpose and value; about uniqueness *vis-à-vis* what it means to belong to a group; about understanding. In short, it can – if you'll let it – get you thinking about what it means to be human. It can get you thinking about what it means to be you – and how wondrously special an experience that actually is.

The Gift of Self-acceptance

As Mouse and I finish our ride, we seem to share a satisfaction in what we have just done – and in simply being together. He puts me in touch with who I am and invites me to be who I could be: I try to do the same for him.

Somehow the horse knows that you are out there waiting for yourself. The horse will help the man. I began to feel that in the end this is the only job worthy of the horse's effort: not to serve a man's childish egocentric impulses and insecure desires – these jobs demean the horse and the lessons of the mythology. The horse, it seems, has always been there to help carry the man to his own real self, if the rider wants that challenge.

Paul Belasik, *Exploring Dressage Technique*, p.136

The Gift of Self-acceptance

Chapter Eight

The Chance of Alignment

Alignment is a key to the success and longevity of effective individuals, teams and organizations.

Robert Dilts, *From Coach to Awakener*, p.144

To be in alignment is to enjoy the kind of harmony that goes with being 'all of a piece': physical alignment makes it easy to function effectively; psychological alignment means that we can be comfortably at one with ourselves; emotional alignment means that we can be clearer in our interaction with others. In this chapter I want to explore what it means to be personally aligned, and to show how horses can help us align and, if necessary, realign ourselves.

Accepting ourselves: the foundation of alignment

Accepting our own humanity is both a challenge and an essential first step in the process of becoming more self-aligned, because it opens up the way for investigating and understanding what contributes to alignment, or its disruption. When you take yourself on board, it becomes possible for you to benefit from some or all of the following:

- Spotting your own patterns and recognizing them when they occur. This enables you to reliably repeat effective patterns and to begin changing less effective ones.

- Distinguishing the patterns that belong to the other being – horse or person – you are relating to. This allows you to deepen your understanding of where they are coming from and how their behaviour makes sense in their terms – and so allows you to become more subtle and strategic in your attempts to connect with and influence them.

- Understanding how you may be limiting the relationship and what it can achieve by 'pushing each other's buttons'. Long-established habits may still have a current function, in which case you need to understand what it is and why it has become automatic; but they may also be 'historical replays' which now have little or no current value. In this latter case, better understanding of the short-circuit between stimulus and response should give you a better chance of working out how to make the pattern less of a knee-jerk response so that you can steer towards different actions and different outcomes.

- Taking credit for your own thinking and actions when they are effective, and recognizing your own 'recipes for success'. Doing the right (or effective) thing instinctively is great – but if you don't know how you achieved what you did you're not in a position to repeat it. Itemizing your own step-by-step processes gives you a warm sense of self-recognition (emotional credit when you need it for new or difficult challenges!) and also allows you to replicate your successes.

- Giving credit to another (horse or person) when it's their due. When you give credit to others you enhance their sense of well-being and self-worth – and through this act of generosity you also grow and become more secure in your own identity.

- Accepting when you have made a mistake – and being charitable to yourself rather than critical or undermining. The most influential words we hear are usually those we speak to ourselves in the privacy of our own heads! Accepting yourself when you are at fault allows you to move on instead of becoming blocked by recrimination and self-doubt.

- Showing charity, and even empathy, when someone else makes a mistake. Reminding yourself that this was not their deliberate intention. Blame – even on those rare occasions when it is actually justified – sours relationships and blocks communication. Assuming good intent on someone else's part gets you investigating how what they said or did could have made sense from their perspective. Once you have stepped imaginatively into their shoes (taken second position with them), your own first-person sense of outrage, anger or disappointment is lessened and you have opened up the way to search for strategies and solutions.

Alignment

As some of the examples in the previous chapter showed, it's only when we are willing to separate out 'their stuff' from 'our stuff', and to take ownership of what's ours, that we can begin to recognize and analyse the patterns of our thinking, feeling and behaviour, identify where they disturb our personal alignment and cause us problems, and take action to improve things.

It's part of being human, of course, to get out of alignment and seek ways to realign. We naturally monitor this unconsciously as well as consciously, perhaps most obviously at a physical level. When I recently twisted my knee, partially dislodging my patella without realizing it, my calf muscles began to ache intolerably. After about twenty-four hours of pain and puzzlement and wondering whether I was perhaps suffering from a deep-vein thrombosis of unknown origin, I found myself standing on my 'good' leg and compulsively shaking and flicking the sore one. After a few minutes of that I sat down again – only to find that the discomfort had virtually gone. Immediately I 'just knew' that whatever had been wrong was now righted. I later understood that the instinctive shaking movement had substantially realigned my knee joint! (This was confirmed by my physiotherapist.)

When we experience a dislocation – even quite a small one – that is emotional or mental rather than physical, we can also feel a sense of wrongness and discomfort. We may find ourselves worrying away at a thought, or going over and over an incident, or wishing something hadn't happened, or that something could... These mental processes are equivalent to the physical shaking movements that began to realign my leg. We may not know as instinctively how to realign ourselves, though. Sometimes we can become aware of the misalignment as we blurt out our feelings, or try more logically to think our way through the problem. Sometimes we may be able to 'shake ourselves back together' on our own, though at other times we may need outside help and perhaps additional support in order to complete the process. In any case, an understanding of just what is involved, and on what level of our being we are at odds with ourselves, will make the process of re-harmonization quicker, less chancy and probably more lasting.

This is a moment of celebration – but it's more than just that. It's clear from their body language that Gaz and his young stallion are both comfortable in their skins: they are fine with who and what they are; they are aligned.

Just what is involved in alignment?

The concept of 'personal alignment' implies that a number of things should 'run alongside' each other, and that currently they either are, or are not, doing so. The word alignment, of itself, invites us to spell out just *what* is being brought into line with *what*. We may 'know in our bones' when we are *at one* and *all of a piece* with ourselves, as common everyday wisdom expresses it, but understanding, as opposed to simply

The Gift of Integrity

sensing, is also necessary if we are to be able to create and recreate that sense of oneness rather than just experiencing it randomly.

The Neurological Levels

A tool which I have found to be both straightforward and powerful in helping myself and others develop this increased self-harmony and self-influence is the NLP model of the 'Neurological Levels'. When the early developers of NLP started to investigate in depth just exactly how people thought about themselves and the world, they found that some form of hierarchical mental categorizing was being used, and that this seemed to be both innate and quite unconscious – hence the use of the word 'Neuro-logical' to imply that systematic categorizing is an inbuilt human faculty.

Mental categorizing seemed to be fundamental to the way human beings made sense of their experience, both external and internal. Having a system of categories allows us to make judgements, for example, as to what is more and what is less significant, what is more or less personal in its impact on us, and to make decisions about what is involved in managing ourselves in relation to our physical environment, other people and the principles according to which we seek to live. By making these judgements, big and small, about many things each and every day, we are able to navigate through events and options, gaining as we do so a sense of meaning and personal significance. This ability to assess things relative to each other, both externally and as they occur within our own minds, is at the core of what it means to be human. It tells us *how* Descartes' assertion '*I think, therefore I am*' actually operates in practice.

So just what kinds of categories are involved, and how are they organized? The concept of the Neurological Levels (nowadays often referred to more simply – but less clearly – as the Logical Levels) was taken up by Robert Dilts from the anthropological work of Gregory Bateson and, through Dilts' work, it has become one of the core models in NLP. Dilts was one of the early developers of NLP and has continued to be one of the most significant thinkers and teachers within the field. This is Dilts' explanation of what the levels involve and how they function:

Gregory Bateson pointed out that in the processes of learning, change and communication there were natural hierarchies of classification. The function of each level was to organize the information on the level below it, and the rules for changing something on one level were different from those for changing a lower level. Changing something on a lower level could, but would not necessarily, affect the upper levels; but changing something in the upper levels would necessarily change things on the lower levels in order to support the higher level change. *Bateson noted that it was the confusion of logical levels that often created problems* [My emphasis].

Robert Dilts, *Changing Belief Systems with NLP*, p.209

The relationship between the different levels is usually shown in one of two diagrams: one looks like a pyramid and the other like the set of concentric circles that form a target.

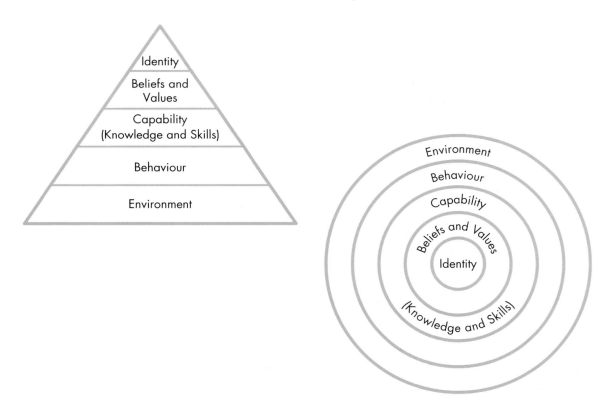

When I drew these diagrams out for one of my coaching clients, he immediately commented: 'I see that as a set of stacking discs'. He had conflated the two flat diagrams into a single three-dimensional one, which seems to me to be even clearer and more useful.

Bearing in mind that each level is controlled by the level(s) above it, let's look at what each comprises.

Disc 5 (bottom disc: **Environment**. It is easy to cue what's involved by asking yourself the questions: '*Where, when and with whom?*' It includes all those things that are external to us, such as surroundings, equipment, arrangements, time and place, who else is involved. In horsy terms, an example might be: *a half-hour lesson in the school on a Friday afternoon, shared with two other pupils.*

Disc 4: **Behaviour**. Cue question: '*What?*' Involves actions. A horsy example might be *jumping*.

Disc 3: **Capability (Knowledge and Skills)**. Cue question: '*How?*' Examples: *losing balance on landing over a fence; falling out through the shoulder; seeing a stride.*

Disc 2: **Beliefs and Values**. Cue question: '*Why?*' Example: *thinking it's best for your horse to go unshod, unclipped and to live out because 'it's more natural'.*

Disc 1: **Identity**. Cue question: '*Who?*' Example: *I am a happy hacker; that horse is an Irish Sports Horse.*

When our **behaviour** within our **environment** is within our **capabilities** and fits with our **beliefs and values** and our sense of **who we are**, we are aligned. We are likely to feel comfortable or at one with ourselves. When any of these ceases to fit with the others, we will feel more or less uncomfortable, disturbed or even distressed. Here are some examples:

- When we feel over-faced by the size of a fence (*conflict between capability and belief – i.e. 'Don't think we can manage that'*).

- When someone we respect tells us to discipline our horse more strictly than we'd like (*our beliefs versus their beliefs*).

- When we know what we should be doing but not how to do it, e.g. trying to perform a lateral movement such as shoulder-in or travers but being unsure what specific aiding to use (*issue of capability*).

- When friends try to persuade us to do something (e.g. compete or go on a sponsored cross-country ride) that 'doesn't feel like me' (*behaviour versus identity*).

As these examples show, alignment can be compromised on one level alone or by a conflict between two or more levels. For so long as something is 'out' we will feel uncomfortable. The steps towards restoring comfort and alignment are to:

1. Identify just which Neurological Level or levels are involved.

2. Then ask yourself which is the easier level to change.

Initiating change is usually easier on the lower levels and harder on the higher ones. For example, if you have become convinced that your horse would be better going barefoot you may find it very difficult to convince a farrier who is wedded to shoeing (an issue of beliefs and values) to go along with you. His *beliefs*, and probably his sense of professional *identity* as a farrier, will both be invested in continuing to shoe your horse. It may be better instead to respect that your beliefs and his both have an important personal meaning, and instead make a change on the lower level of *environment* by simply changing your farrier to one who shares your belief about the value and appropriateness of barefoot work for your horse. The change itself is *environmental* (different farrier) but it is supported by the fact that the new farrier's *beliefs and values*, as well as his professional identity, support the outcome you want.

When you really can't bypass issues of beliefs and values or identity you will probably need to find some way to make your adjustments on these levels themselves, since they are high in significance and exert powerful control over the levels beneath them. In the case of beliefs, for example, you may be able to help someone (yourself or someone else) change a belief that is very specific by appealing to one that is more general, especially if you both share it. That may then open up ways of approaching the situation that you hadn't seen before.

At the time I was doing my training in NLP I was distressed about my relationships with my horse and my trainer. I couldn't seem to get my horse working on the bit unless I allowed him to work in a long, deep outline. My trainer thought he was ready to progress beyond this, and that he should now be asked to carry himself higher and rounder. I could see that he found the higher outline difficult because, when he was asked to assume it, he stopped swinging through the back and tended to go against the hand; but whereas my trainer could get him to work in the more advanced outline, I couldn't. It's a version of the same problem that we have been experiencing more recently with Mouse; and I know of many riders who have been through similar experiences, over a variety of training issues, with their trainers and horses.

We all need help from outside at times, and sometimes managing the triangle of relationships between rider, horse and trainer can be difficult. Trainers have a professional agenda based on their experience, and more highly developed skills; pupils can only do what they can do, and however conscientious and able they may be, they are on a different place on the learning curve from their teachers – and horses can easily get confused somewhere in the middle! As a result of such differences in beliefs, skills, expectations and behaviour, any partner in the triangle can quite easily feel out of alignment. Trainer, pupil or horse can feel anxious, depressed, angry, puzzled, helpless, as I did.

Because I was learning a whole range of new NLP skills at the time, and practising them by working with other course members, I found

The Gift of Integrity

myself bringing up my horse-related problems and dilemmas as material to work on.

It was clear at the time that my horse was the focus of many of these problems: looking back years later I can now see that he also helped me by making me aware that I was out of alignment – surfacing my distress to the point where I needed to reflect more deeply and act more congruently. Horses seek clarity and consistency from their riders: only these qualities give them the information about their place in the herd that makes them feel secure. Torn between our trainer and me, my horse was puzzled and conflicted, as his physical behaviour, mental confusion and emotional state all showed. Because these conflicting requests made him insecure, he confronted me with the need for thought and action – about myself as much as about him – and succeeded in highlighting questions that turned out to be crucial to the development of my independence and confidence as a rider.

Ollie's face and ears tell us that he is calmly attentive – but also slightly puzzled. I have not been on his back for over a year, and I think Leo has caught the moment when Ollie is waiting for me to tell him who I am and what I want. He's looking for guidance and clarity.

At that time I was distressed by the conflict between what I believed we should be expecting of my horse, and what my trainer expected of him, and we were all handicapped by the discrepancy between our differing evels of skill. I believed then that if your hands were gentle horses would automatically work into them: if the horse resisted, I thought this was a sign that you were asking too much, or too much too soon. I give myself credit for idealism here – but I didn't understand that however warm riders' hearts are they still need to set clear expectations, and that whenever an individual (human or horse) is asked to step up to a new level they may, in the short term, become awkward, uneasy or even unhappy.

There was the further problem that I didn't have enough confidence in myself as a rider (an issue of *identity*) to insist that my trainer and I both had to set our expectations according to the limits of *my* skills rather than *hers*. I can now see that no rider can ask their horse effectively for things they themselves do not have the skill (*capability*) to help him achieve.

My distress – and the tensions within the triangular training relationship at the time – were caused by issues involving *misalignment*. Our problems centred on the highest logical levels – *identity*, *beliefs* and *capability*. I now think that finding a common understanding on the level of capability would have given my trainer and me the leverage for change and resolution that we needed.

I believe that the real priority in a situation like this is to act in alignment with oneself: you should respect your own sense of internal alignment or dislocation. Only with internal alignment can you remain authentic to yourself and act consistently towards others. It's clear to me now that my trainer and I both wanted the best for my horse and for his athletic development (shared *values*). We would probably also have agreed, if we had discussed the issue in these terms, that he would find things clearer if we gave him the same message (shared *belief*). Making that possible, however, was going to require us to be consistent in our *behaviour* towards him, which would mean that we could only ask of the horse what both of us were capable of following through. I realize now that this would probably have required a change in our ambitions for the pace of the horse's development – this could have restored alignment while still honouring the good faith and shared hopes that began our training arrangement.

The Gift of Integrity

Resolving problems of interpersonal misalignment

This problem is one that often bedevils the rider/trainer relationship, and I think many people stop having lessons because it's one that's hard to confront openly. Horses show their misalignment through difficult or confused behaviour: human beings often attempt to hide it or flee from it by leaving the situation rather than identifying and attempting to address the root cause. Both parties are likely to be afraid of hurting or offending the other if they are the one to raise the issue of their dissatisfaction. But if they don't, someone – perhaps both, perhaps all three if you include the horse – may feel the discomfort, even distress, that comes from misalignment.

If one is to resolve situations like these, it's important first of all to assume that the other person has the best of intentions. A trainer wants his or her pupils to improve and to help their horses learn and perform well. A pupil/horse-owner wants to do the best they can and to learn as well as they can. This needs saying out loud by one or both, as it lays a respectful foundation for any further negotiation and possible further progress. From this point on, though it's the trainer's responsibility to stretch the rider to progress, nonetheless the rider's skill level at any time has to be what determines the rider/horse partnership's rate of progress. As de Kunffy reminds us in his workshops, '*always you begin with the rider*'. For a trainer to remain personally aligned in working to the rider's limitations rather than to his or her own ability to improve the horse, he or she needs to prioritize their role as *teacher of the rider* over the alternative role as *trainer of the horse*.

Hindsight is a fine thing: if I had been as fully able in those earlier days to use the Neurological Levels model as I am now, I might have been able to help all three of us out of a situation that was probably frustrating and unproductive for us all. Even when you have the understanding, making the changes in your own thinking and feeling (and having them sufficiently 'in the muscle' for them to affect how you behave towards others) can take time. I know at first hand just how much the Neurological Levels model has helped me in managing myself and others better. Using it has meant that I have stopped feeling so inadequate as a horsewoman and I now get less frustrated with my horse. My fragmented personal alignment has recovered and because of that I have felt empowered to make some important and, at times, difficult decisions.

How horses help people with alignment issues

As this personal example has shown, for riders it's often a discomfort around our horses that shows up where we're out of alignment; but it can frequently also be the horses that shake us back in again.

How do they do this? As I've already explained, it's clear from many people's experience that horses mirror us to ourselves and show us places where we are at odds within. Horses notice even the slightest discrepancies or mismatches between what we say and what we do, what we do and what we feel, what we aim for and what we fear – not because they understand what's affecting human beings in human terms but because they unerringly recognize lack of alignment at any level. Horses are the touchstones of our personal truth – or of our failure to keep that faith with ourselves. One horse facilitating on the Acorns2Oaks leadership course that Leo and I attended kept cutting in towards a participant who was trying to lunge him, because (as the would-be lunger admitted when questioned) in his heart that was what the lunger really wanted.

Horses are not fooled by our words: they believe our real feelings rather than those we're trying to convince ourselves of. Where we are in conflict with ourselves they may be puzzled, frustrated or even annoyed – *because we aren't coming across clearly to them.* They sense discrepancies but are at ease with wholeness. They effectively monitor and act as a clear outward barometer of our degree of personal alignment.

If we are willing to grapple with the issues they so unerringly highlight for us, horses also offer us the possibility of a dialogue through which we can begin to explore those issues and realign ourselves. Such a dialogue will involve us and the horse – but inevitably it will also involve us engaging with our inner shadows. A horse will sense fearfulness or lack of confidence and respond to it however much we attempt to deny it. But if we are prepared to accept that we are fearful and take the pressure off, a horse is much more likely to go along with us step by step in pushing our own boundaries, effectively helping us to become more courageous. Many of the people who responded to my questionnaire said that their horses and ponies had helped them when they were sad or

The Gift of Integrity

stressed – just by being there. Their horses accepted the individuals and their feelings for what they were, which made it easier for these people to accept themselves.

'I feel peace just being in their presence.'

'I have gone through a very stressful and emotional time and the only thing which has seen me through this horrific experience has been my horse, who has been a brick for me… Interestingly the days I have not ridden I have ended up being stressed out and tearful and not being able to cope.'

Personal congruence – the core of the helper

How do horses bring someone peace or 'see someone through'? I believe that their healing influence stems from their own integrity, and I want to suggest that, like the best of human therapists, they possess their own individual *congruence* which can powerfully affect those who come into contact with them.

There is a strong similarity between the concepts of congruence and alignment, though I shall use them separately in acknowledgement of their different origins. Writing in 1961, fifteen years before the earliest publications in NLP, Carl Rogers, the American psychotherapist and founder of what became known as 'person-centred therapy', said that, in order for therapeutic change to take place, it was essential that the therapist:

'…be, in the relationship, a unified, or integrated, or congruent person. What I mean is that within the relationship he is exactly what he is – not a facade, or a role, or a pretense… Until this congruence is present to a considerable degree it is unlikely that significant learning can occur. Though this concept of congruence is actually a complex one, I believe that all of us recognize it in an intuitive and commonsense way in individuals with whom we deal. With one individual we recognize that he not only means exactly what he says, but that his deepest feelings also match what he is expressing. Thus whether he is angry or affectionate or ashamed or enthusiastic, we sense that he is the same at all levels… We say of such a person that we know 'exactly where he stands'.

Carl Rogers, *On Becoming a Person*, p.283

In replying to my questionnaire, David Harris, who is a trained NLP Master Practitioner and Coach and who now runs Acorns2Oaks equine-assisted leadership training courses, made it clear that, in his experience, because of their own congruence horses have the power to help us become more so:

'*They teach me patience, congruence, how to breathe, how to be fully present, what leadership suits me best. I feel I have learnt as much from my horses as I have done on all the trainings I have ever taken part in.*'

If we go back to Rogers again, we can see how the congruence that horses themselves possess naturally can cascade outwards through their relationships with human beings and have such potently transforming and enabling effects:

> The greater the congruence of experience, awareness and communica-tion on the part of one individual, the more the ensuing relationship will involve: a tendency toward *reciprocal communication* with a quality of increasing *congruence*; a tendency toward more *mutually accurate*

For me this ride was a special experience: the fact that I could ride Ollie 'on the weight of the rein' told me that even though we were unfamiliar partners for each other we were able to accept both ourselves and each other for who we were. There's a trust in ourselves and in each other which felt great at the time and which I think comes across in the mutually thoughtful collaboration of this moment.

The Gift of Integrity

understanding of the communications; improved *psychological adjustment* and functioning in both parties; *mutual satisfaction* in the relationship.

Carl Rogers, *On Becoming a Person*, pp.344–5

In the case of therapy, it's customary to expect the therapist to play this role: where riders experience healing or growth through their riding the 'one individual' whose way of being they recognize as initiating therapeutic change is often the horse. Because horses are of a piece with themselves, what they experience, what they notice and what they communicate are in synch with each other: their communication is direct and unambiguous. More than this, though, they offer us a model of *how it's possible for us to be*, within ourselves and with others. They show us, in other words, how congruence is 'done'. There is a sense of clarity and, indeed, often of cleansing in being around them.

I am riding in the school. Getting to know Mouse over the past eighteen months has been like getting to know a person: it has taken time. Physically, he has needed to become looser and build more muscle over his top-line – letting go and building up can't be hastened. That, we expected. Emotionally, we were less prepared for hiccups in building our relationship with him, and more distressed by them. Yet his reluctance to trust us when something alarms him out hacking is really no more than just that: effectively he's telling us: '*I don't entirely trust you yet, so in the pinch I shall continue to look after myself rather than taking your word for it.*' Since Leo and I both pride ourselves on being trustworthy, this has been a hard pill to swallow. Recently, however, he and we seem to be making more headway with each other. I am becoming clearer about what I want and no longer confusing Mouse with my tentativeness. Those defining exchanges about going past the barn (see Chapter 5) have hugely increased my confidence – and his in me. For his part, Leo has discovered out hacking that a moment of stillness and acceptance when Mouse hesitates, followed up with a low growl of clear yet non-threatening reinforcement, is much more effective in getting Mouse to continue past whatever is alarming him than an old-fashioned boot in the ribs.

Today, as we work in the school, this new-found clarity of communication and mutual understanding means that Mouse and I are really concentrating and flowing with each other. For almost the first time in my years as a rider I truly feel what it means to 'ride a horse from behind and receive him elastically in my hand'. It's amazing. It's not consistent, of course: it's like a conversation between friends that from time to time takes off from mundane exchanges into spheres of intimacy, speculation, confidence, ease… It feels precious.

We finish our work in the school and come back up the track. As we approach the barn Mouse starts to duck off the track towards it. I growl and put my legs on – and without further hesitation he marches cheerfully past and up the track into the yard by the house, where we turn round and come back again…

Mouse's role in my realignment

During the time we have been building our relationship, Mouse has confronted me with my tentativeness, my anxiety, and my self-doubt. In doing so, he has pinpointed my areas of non-alignment: my *beliefs* haven't matched up to my real *capability*, and have therefore adversely affected my *behaviour*. Thus our developing partnership has given me the opportunity of engaging with myself as much as with him. And, of course, he has given me the purest and clearest confirming feedback of the smallest changes I have been able to make towards realignment. He has nudged me into starting a process that has become more than a simple bringing into line: creating a spiral of reinforcement, he has set off changes in me which bring about changed reflections from him, and so further changes in me, and further changes in our relationship.

Let's just spell that out in more detail.

- Mouse has challenged me at a behavioural level.

- Which has made me aware that I have often been performing below my real level of skill (*capability*) because of my rather dented beliefs about my adequacy in this area.

- Mouse has enough of the same training needs as our former horse to give me the opportunity of working with him (and so, with the 'unfinished business' from that earlier relationship) in a way that feels aligned and personally congruent. (I think of my current trainer as an

The Gift of Integrity

external mentor and consultant rather than as an ongoing 'third partner' in my relationship with Mouse.)

- Through additional years of judging and of observing great trainers, I have more confidence in the appropriateness of my judgement about what I am trying to do, and am more assured that my ability to do it is 'good enough' – for Mouse, for Leo and for me.

It seems to me, taking a third-position perspective on our situation, that these changes between them encompass behaviour, environment and capability, and are adding up to changes in my beliefs about my identity… I begin to have hopes of myself after all.

The Gift of Integrity

Chapter Nine

A Way of Being

The rider must take care not to be tempted into believing that art is only produced when doing the advanced movements. Art can exist at any level…

To be an artist is to answer to an irresistible calling – a vocation based on a sense of self-conviction… It is a dedication of one's life to the simplicity of the daily task…

Erik Herbermann, *Dressage Formula*, pp.156–7 (1980 edn)

On those occasions when I wonder whether I am too idealistic – too romantic, even – about riding, I remind myself that others with longer experience and much greater expertise have also described the harmony (however momentary) between horse and rider as a meditative, even spiritual experience. Herbermann, de Kunffy and Belasik are all supreme technicians and pragmatists – yet for each of them the best analogy for riding at its harmonious best is the unified, apparently simple, focus that characterizes artistic creation. I say 'apparently simple' because such a simplicity is the result of processes that are infinitely complex. It is the flowing together of these processes that make the experience *seem* simple, spare and 'just right'.

> Nature is art's raw material. Art is invested with human imagination, a longing for beauty, with ecstasy in understanding, the thrill of intelligent planning and a profound sense of fulfillment when something new is born.

> Riding is…a Baroque art, with the goal to develop the natural, inborn talents of each individual horse… Baroque art with its swirling, elaborate, animated, yet ordered turbulence, comes to its finest fulfillment when a fully schooled horse is allowed to display his beauty.

> Charles de Kunffy, *The Ethics and Passions of Dressage*, pp.10–11

These thoughts on art are underwritten by Belasik's observation: 'It is art that can't stand still. It is art that is alive. Art that must be watered and fed. In the last analysis it is really the art of living.' (*Riding Towards the Light*, p.11)

I remind myself, too, that such experiences are not the sole prerogative of the advanced and the highly skilled. Someone who filled in my questionnaire said that horses enabled her to find 'something I am OK at that I love and that can make me cry with joy'. I am not particularly experienced at jumping, but going over a small course of jumps in a field on one of my friend Gaz's cobs – a horse who loved jumping and knew exactly what to do – made me feel that everything was truly right with the world. I often felt the same hacking round the farm on a summer evening on our old friend, Lolly.

I am in my early twenties again. I am riding on the Sussex Downs with a colleague from work. The little black mare I have hired from the riding stable is a whiz with gates – opening and closing them is like a passage from a ballet. We climb up the steep inner slope of the Downs onto their high ridges and folds – and suddenly the skies open and within minutes rain has soaked us to the skin. There is no point in turning back, so we go on. Now there is thunder and lightning. Cattle loom through the sheeting rain as though glimpsed through net curtains – solid yet dim. Rain rises from each hoof-fall. For some reason the horses are not frightened but, like us, exhilarated. We are cantering along the ridge, along the roof of the world. Our shouted conversation is as intermittent yet as companionable as the cattle sporadically appearing through the mist. Only when our path reaches its crest and approaches the downward slope again is there any change: then the clouds and the sheeting rain seem all at once switched off, and the sun takes their place. We ride back to warm baths and a roast dinner.

The Gift of Transcendence

The beginning of a ride; the beginning of marking out a pattern; where are horse and rider headed?

The shaping of energies

In dressage, we often talk about a horse's 'way of going', by which we mean the totality of his physical, mental and emotional being at any point in time. In NLP, this same totality of experience is described as a 'state': the word encompasses a similar comprehensiveness of mind-body experience. In this chapter I want to celebrate this process of creativity as one that is not just one-way but *two-way*. De Kunffy rightly emphasizes that 'one cannot shape a horse, only his energy': this sense of respecting and directing what is already inherent in our equine partner has reciprocal effects on us, too. In shaping his energies we can, if we are prepared to, allow him to shape ours in return.

How does this come about? In previous chapters I have tried to show that horses:

- Reflect us to ourselves without judging us.

- Focus our attention without coercing us.

A 'going with' moment with our old friend Lolly.

- Teach us without being dogmatic.
- Invite and support our authenticity through their own congruence.

They can also:

- Learn without subservience.
- Show sensitivity and apparent sympathy towards emotional trauma, distress and disturbance.
- Have a calming effect on human beings through their slower physiological rhythms as well as their solid, warm physical presence.

By these means they shape our energies as much as we shape theirs.

The Gift of Transcendence

A special relationship and a singular state

Of course we can also feel fearful, frustrated, upset, disappointed and even rejected when we are around horses. Nonetheless, it was clear to me when I very first asked myself, 'What do horses do for us?' that the answer had to be 'something pretty important' since so many people remain passionate about riding and caring for them in spite of the negative elements in their riding lives. What is it that sustains us when fear, cost, setbacks and injuries confront us?

First, it's *the relationship* that challenges, taxes, enriches and improves us, provided we are willing to engage with it and learn from it honestly and attentively. One of the respondents to my questionnaire said this:

'Horses have been my constant. They give unconditional love. When I was a teenager my pony was my escape from the stress of school and peer pressure. My pony taught me to be responsible, sensible and caring. Being around other horsy people taught me how to relate to people of all ages. But most of all, it was Whistles (my pony) I ran to if I felt upset or sad. Horses give me two distinct aspects of my life – riding is my sport and passion, and my horses are my best friends. I think owning horses has made me analyse how to communicate non-verbally and this is a transferable skill with adults and children too. Trying to work out why a horse behaves in a certain way is fascinating and has allowed me to become more questioning, look at my own behaviour and, most of all, listen. Working out a horse's behaviour pattern associated with stressful activities and assisting in changing that behaviour is incredibly rewarding. Horses have also given me the most thrilling experiences of my life.'

Second, it's the memory and the promise of that singular state we can experience in partnering them: whether we think of it as synchronicity, flow, artistic creation, oneness, being ineffably present… Belasik (*Riding Towards the Light*, p.12) calls the dressage arena 'the perfect Zen sand garden'. In De Kunffy's experience, riders with humility and attentiveness find that 'their powers of concentration…deepen to a meditative state, oblivious of anything outside the harmonious absorption of their communication with their horse.' (*The Ethics and Passions of Dressage*, p.12) In *Dressage Formula* (p.12) Herbermann argues that the effort to become proficient in horsemanship can lead us to moments when we

The Gift of Transcendence

'experience a sublime state, an ecstasy that sets our physical and spiritual senses alight with joy.' What they are all talking about, I believe, is the kind of experience that results from a special way of being.

The transcendence of 'going with'

Mark Rashid reminds us that the aikido precept of 'going with' yourself and the horse as you progress through the essential steps of learning *at your own pace* is what allows you to be happy and at peace:

> For me, one of the best parts of working with horses is learning how to go through the steps. It's what gives me the understanding I need to do the work well. I feel that a big part of being able to learn the steps starts with being truly happy with where you are in your own learning or level of expertise. In other words, it's about learning how to 'go with' your personal learning process and knowing that process will be different for each person.
>
> Now, when I say you should try to be happy with where you are in your learning, that doesn't mean you need to be satisfied. Being happy just means you accept (or 'go with') where you're at any particular time. Once you learn how to accept where you are, you're able to take in as much of that level as it has to offer. And *that* is what will ultimately move you on to the next level. If you don't accept where you are in your learning, you will constantly be looking past where you are and ultimately miss most of what you need to learn while you're there.

Mark Rashid, *Horsemanship Through Life*, pp.96–7

This is very different from the 'driven' feeling that riders can experience either as a result of competitiveness with others or of being over-urgent with themselves, or over-perfectionist. This 'drivenness' is actually the enemy of learning, because it imposes a filter of 'oughts', 'musts' and 'shoulds' between you and what is happening in yourself, in your equine partner and between the two of you. It is also the enemy of peace, because it allows you no rest: however much you achieve, there will always be another goal to strive for. Rashid's argument is quite subtle: riders do need goals and aims – but they need to use them as fluid guidelines, not as rigid tramlines. When something you ask for doesn't happen, or when circumstances turn out differently from what you'd expected, you need to check whether it's really appropriate to continue pursuing your goal this minute or whether, in Rashid's words 'it didn't need to be today'. For the greater goal is to develop and maintain a seamless flow of energy between you and your horse: without that, there's a real danger that what you do will be mechanical and what you achieve will be without grace.

> Over-reliance on hands, legs, and seat can lead to a mechanical feel from the horse. While the movements and responses coming from the horse may be all there, they may lack a smooth flow from one action to the next...
>
> As I studied the proper use of the centre of power in aikido, I began to look at that hesitation between horse and rider differently. When performing a technique in aikido with a partner, the goal is to have a flow of seamless energy from the time the person makes contact with his partner, until the technique is completed. There should be no hesitations, no glitches, no stopping, and really not even any thought involved – just a smooth flow of energy from one spot to the next between two moving individuals.

Mark Rashid, *Horsemanship Through Life*, p.144

Rashid is talking here about flow in aikido and in horsemanship: the implication, however, is much wider. When our horses invite us into their 'Zen garden' (whether it's a dressage arena, a jumping arena or the fields and paths along which we hack with them), and when we have the love and the wit to accept their invitation, we open the way to that flow of energy, that presence of all of ourselves in the moment, which constitutes the most precious peak experiences in life and which give us what the psychologist Abraham Maslow referred to as 'transcendence'.

> Transcendence refers to the very highest and most inclusive or holistic levels of human consciousness, behaving and relating, as ends rather than as means, to oneself, to significant others, to human beings in general, to other species, to nature and to the cosmos.

Abraham Maslow, *The Farther Reaches of Human Nature*, pp.291–2

The Gift of Transcendence

It seems to me that once, as riders, we have experienced such a moment of transcendence, we too cannot un-sense it: it is truly transformational. One rider said: 'Riding can be a good spiritual exercise: the biggest lesson I learn every day from my horse is to SLOW DOWN.' The equine healer and writer Margrit Coates replied in a similar way when she said: 'Horses have taught me to be centred and to create stillness at my core'. What these people are articulating is an understanding that, once discovered, becomes a benchmark. Of course the spiritual rider can be mundane, the self-respecting slowness can be ruffled, the person temporarily dislodged from their centre. This is part of being human. What is significant is that, once you have felt your spirituality, your calmness, your 'centredness', it becomes an internal benchmark by which to judge your ongoing experience, and a lighthouse by which to steer yourself back.

An altered state

Transcendence is not about leaving things behind: rather, it's about being so intensely *in* them that you cannot but be drawn into larger and larger awareness. Transcendence is about *experience* and *context*. T.S. Eliot called it: '...the intersection of the timeless moment with time'. Linking this duality with the concept of 'slowing down', Herbermann makes the observation in *Dressage Formula* (p.41) that: '*Our reactions must be lightning-quick, but our attitude towards the horse should always be as though we had all the time in the world. It is a calmness that seems at odds with the speed of reaction.*'

I think it is because riding can involve every part of ourselves – body, mind and emotions – so intensely, that it has the power to take us both deeper into and at the same time beyond ourselves. What are the characteristics of such a state?

- A calm yet focused attentiveness.

- A readiness to act and react – as though all our nerves, muscles and thoughts were available and 'online'.

- A duality of mindfulness – acute simultaneous awareness of self and other.

Calm yet deeply attentive. Nikki's eyes are defocused and her facial muscles relaxed. She is 'in the zone'.

- An ability to interact with here-and-now detail yet with full reference to its longer, larger context and consequences (e.g. aims and implications).

- Right-brain feeling and creativity working hand-in-hand with left-brain logic and analysis.

When I enumerate these characteristics, I find I am describing an altered state of awareness akin to some of the healing and creative hypnotic trances that I used for many years as a therapist to help clients heal and

The Gift of Transcendence

grow. Riding can indeed be its own active trance state, with a similar potential for releasing, healing and enhancing. If we look at the process of formal trance-work we can see that it has similar ingredients and a similar structure to the process I have just described. In essence, there are four elements in a specific sequence:

1. Focus external awareness intently to the point where extraneous stimuli are filtered out.

2. Intrigue the mind with new possibilities or by presenting familiar ideas and experiences in an unfamiliar light.

3. Pose questions in such a way that they cannot be answered fully by the conscious part of the brain, but require the deeper knowledge and the making of creative connections that can only happen at an unconscious level.

4. Re-orientate to current experience having integrated new understandings in a deep and lasting way.

Using the altered state

Step 1: Focus

It is clear that, when we allow our horses to sharpen our focus on specifics with all our senses 'online', the potential exists for entering an altered state. Of course we are not going to be closing our eyes and becoming deeply relaxed as we might in a therapy room. Nonetheless, if you observe riders in that meditative state you will see that their facial muscles have become fuller and differently toned, and that their eyes are defocusing rather than outwardly focused. Intent focus leads on to soft focus (the pupils of the eye defocus, so vision becomes wider and blurrier and more use can be made of peripheral vision); outward concentration leads on to inner absorption; the senses monitor experience not with intent conscious awareness, but with unconscious effectiveness (as quite frequently happens in familiar or repetitive actions such as jogging or long-distance driving); thought and sensation become fluid, expressive, intuitive, creative.

Roy's inner absorption allows him to concentrate, to monitor, to react and yet to be entirely free of tensions. He is absolutely 'in the moment' – wholly available to his experience and to his horse.

Step 2: Intrigue the mind

As Timothy Gallwey discovered, and explained in his *Inner Game* books, it is relatively easy for a coach to encourage a habit of inquisitiveness in their pupils. Watching the seam of the tennis ball as it moves is a novel approach, which just happens to bypass your conscious 'critical sensor' and enables you to integrate naturally with the flow of the movement rather than consciously trying to predict or control it. For similar reasons, some golfers like to use balls with logos, because looking at the logo adds to the sense of focus on the ball. Becoming curious about what happens when you make a small shift in weight, or alter the angle of your shoulders, or the inclination of your body, as you are riding attunes you to the flow of energetic information between you and your horse rather than involving you in the 'trying' to produce that extra collection, that shoulder-in, that lightness or engagement... Becoming curious about the meaning of someone's changing facial expression, or the way they favour certain metaphors in their everyday speech, keys you in to how they experience the world and allows you to communicate more closely and effortlessly with them. Tried and trusted procedures bring familiar results in life as in riding: curiosity and experimentation are what deliver the new.

> The fact that what is novel and fascinating actually heightens brain activity is a very important, though still generally unappreciated, precondition for all forms of creativity-oriented psychotherapy and mind-body healing experiences.
>
> Ernest Rossi, *The Psychobiology of Mind-Body Healing*, p.28

Step 3: Pose questions for the unconscious

In riding, as in life, we often know what we should do, yet somehow we cannot manage to do it. Conscious exhortation by oneself or others (such as trainers or mentors) to change one's ways rarely produces change. Only by approaching the issue at an unfamiliar level or from an unfamiliar angle can we engage our own unconscious ingenuity and self-knowledge to produce different solutions.

In NLP this process is known as 'reframing': when you change the context of an issue or a problem you change how it is perceived, in much the same way as reframing a picture heightens awareness of different aspects of the image.

I am coaching Roy and his horse, Cori. We have become particularly aware this morning of how Roy tends to hook back with his left hand, and of how this contributes to Cori overbending, thereby making it more difficult for him to engage and carry himself. Reminding Roy to straighten his wrist, which I do a number of times, fails to make a sustained difference. He knows, and I know, and he knows that I know… But the wrist keeps to its old habits. I think about redirecting his attention, and find myself talking about how the great rider Arthur Kottas 'rides through' his elbows: they are bent at 90 degrees and kept close to his sides without any stiffness, and they accommodate the horse's movement with infinite elasticity while continuing to provide a supportive frame. I ramble on about riding behind my friend Gaz out hacking and becoming aware of how similar his elbows were to Kottas's. I quote de Kunffy: 'Without elbows there is no riding'.

These are, of course, suggestions – but they are offered not as commands but as observations. Implicitly, they pose for Roy questions like: 'Could I do that?' 'What would happen if I did?' 'How could I do it?' These are not questions to which his conscious mind can have answers but, because he finds them deeply intriguing, it's inevitable that his unconscious processes will become engaged. They, of course, work far faster than conscious processes, and they transmit directly and almost instantaneously through the molecular mind-body information systems that Rossi helped identify. Within moments Roy's riding – and Cori's way of going – are transformed. Roy's hands are no longer of importance: they are still and perfectly aligned with his forearms. All his riding comes from and through his sensitive, elastic, respectful, supportive elbows. Cori is engaged, 'uphill', soft, able to extend and collect with effortless ease. The elbows have unified horse and rider into one entity.

Step 4: Re-orientate and re-integrate

The conscious mind does have a place in creativity and learning: it helps us understand, mark and replicate just what it is that we have discovered. This is why Gallwey formed the habit (which I follow) of asking his pupils at the end of a coaching session what they would be taking away from it. Articulating the learning allowed it to become integrated at a conscious as well as unconscious level, and made it available for deliberate reference and replication. It also marks a boundary for the altered state and its special work: knowing what you can do with any specific state and how to access it – in riding and in living – is a necessary and valuable skill. The specialization of focus and the altered state it leads to

The Gift of Transcendence

allow us to intervene in our own experience in very precise ways: it's a form of self-applied micro-surgery. Like any adjustment, the test of its efficacy is in the degree to which it improves your life. And for that, the micro-intervention has to be integrated into your way of living and being at a more macro level. It's here that your conscious mind has its place.

Benefits of the altered state

For me, writing this book has been a continuous exercise in exploring what I know – and how I 'know more than I know that I know'. This is the 'what do I take away' part of the process. Reading de Kunffy, Belasik and Herbermann I have a sense that they, too, are documenting the course of inquiries that have taken them from conscious to unconscious processing, from everyday states of riding the best they could to altered states in which, in their horse's company and with his invitation, they entered quite another dimension.

Such a state is a place of living, changing, breathing balance between many things which are often thought of as contrasting, or even opposed. This is not balance as in a static, managed equilibrium, but as an ongoing process of *balancing*. For me, the nearest analogy in my life is how I feel when I am writing – but writing lacks the distinctiveness of being a dialogue that is *shared*. Shared is doubled – at least: it is a very special state indeed.

I've called this chapter 'A Way of Being': I am beginning to feel that I should come clean and substitute 'the' for that 'a'. I do believe that what I have been exploring, in common with my admired mentors and many thinking riders, is a way of being that is, for each of us, analogous to the way of going that characterizes our horse. Beyond that, if I have the courage of my own convictions, I believe that we are exploring '*the* way of being': a holistic pattern of openness to our experiences and a readiness, with our horse's help, to become the anthropologists and philosophers of our own experience, both mundane and spiritual. That is the fine mess – which, thanks to the wonderful ambiguities of language, also carries the meaning of 'an exquisite dish' – which riding gets us into.

I am coaching Leo and Mouse in the school. Leo is struggling to keep

Mouse straight on the track, and to prevent him cutting in on circles. In his efforts to control Mouse's persistent drifting, Leo puts in more physical effort, in the process losing the correctness of his position. I point out to him (nag him about?) the fact that when Mouse veers inwards Leo turns his own body outwards in an effort to correct the error, which in turn allows Mouse to become crooked. Then that has to be corrected. At walk, things are more manageable: at trot Leo is busy correcting the corrections of corrections. He feels frustrated; Mouse seems confused (he tries to do what is asked – is, in fact, actually doing what Leo is mistakenly asking); I feel helpless. Fine coach – fine partner – I am! The worst of it is, we all mean so well.

In his anxiety to get Mouse working correctly, Leo is trying to achieve the impossible: he is attempting to get what he wants through sheer muscle power. As a result, he loses his position and they lose their rapport. Both faces express concern.

The Gift of Transcendence

In desperation I pick on two details. With the convenience of hindsight I have now recall Herbermann's words (*Dressage Formula*, p.19): '*When in trouble: do less! Neutralize. Let the horse settle down and find itself before making new demands.*' Never mind what Leo is trying to achieve or where he wants to go – he needs first of all to regroup and simplify, and to keep his elbows closer to his sides (elbows again) rather than letting the outside one slide forward. That only takes away any consistent supporting frame from Mouse and causes him to fall through his outside shoulder. Keeping his elbows close, though quite soft, allows Leo to do the second thing that's necessary, which is to steer by inclining his hips and shoulders in the direction he wants to go while keeping a consistent downward stretch and support for his mount through his legs. I know that these things need doing; I know Leo can do them; I even know that they should make the difference we are after; what I don't realize is how instant, how magical, an effect they will have.

Different posture; different man; different horse; different partnership. The effort has gone, replaced by harmony. There is still a vestige of carefulness in both partners, but now they are flowing forward together.

The Gift of Transcendence

What had been at times a state of confusion, ungainliness, disharmony, and urgency becomes at once a state of clarity, elegance, harmony and spaciousness. 'Trying' gives way to 'being'. Leo relaxes. Mouse relaxes. They move in close harmony together.

These are technical adjustments, but beneath them lies a change of state. By putting himself in a place of stillness (mental and physical) Leo has created a wholly different set of possibilities for a partnership which – like all our efforts at communicating with each other – is always potentially fragile yet always potentially resilient. It's the way of being that makes the difference. As De Kunffy puts it in *The Ethics and Passions of Dressage* (p.12): 'So can man be elevated by the taming of his horse, through a partnership with him, to become himself the object and the subject of his art.'

The Gift of Transcendence

Chapter Ten

The Exercise of Morality

The horse is our conscience.

Erik Herbermann, A Horseman's Notes, p.55

We make our lives significant, even on this minor planet, which is part of an unimportant solar system in an insignificant galaxy, by asking questions about who we are, what our purpose is, and how to do well. The answers become significant only in the light of another's love and acknowledgement. Whether this other is a human or a horse, the justification of our existence will come from being valued by another.

Charles De Kunffy, The Ethics and Passions of Dressage, p.39

We ride our horses the way we live our lives. Our hopes, our fears, our anxieties, our strengths and our limitations are all played out in stable, field and arena. I believe that, thanks to the direct and pure opportunities our horses offer us, they can teach us much about leading better lives.

What is morality?

In the previous sentence, I chose to use the word 'better' specifically because it is ambiguous: 'Better for whom?' 'Better in what way?' The daily choices we make around horses are essentially ones of morality. In its Latin origins, the word *mores* refers both to the details of established custom and to individual acts of will. Both presuppose the underlying guidance of values or principles. In choosing how we ride, how we feed, how we school, whether we compete, whether we shoe, how we address the challenges and opportunities of developing and disciplining our horses and ourselves, we are exercising choices that are moral in this sense, whether we consciously refer to a belief system or not. The choices we make on a daily basis have moral consequences, for us as well as for our equine friends.

More times than I like to think of, I have dragged myself down to the yard wishing I didn't have to ride. I have wanted to avoid making what feels like a selfish choice. If only the weather would change suddenly… Maybe the sneezes I woke up with in the morning will develop into a cold so I feel too unwell to ride… On such occasions there has been a real sense of duty in my mind, and a real conflict between what I *wanted* to do and what I felt I *ought* to do: underneath, though, I knew that the horse needed exercising, however I felt. If I couldn't do it, someone else would have to. I felt guilty, believing that most people don't have such moments. Of course, I ended up riding, or lungeing, because I had a commitment to my horse that was more compelling than my tiredness, my anxiety, my not-feeling-in-the-mood, or whatever. More often than not I enjoyed myself – and my horse – more than I had expected. I wonder why I put myself through such conflict and such doubt.

I worked with my horse because I have engaged myself to him, to his health, his safety and his well-being. And, in so doing, I have engaged myself to values within me that are more profound than immediate feelings and preferences. I could say the same about my family, of course; but they have greater independence and can speak for themselves. A horse who is shut in a confined space because of my choice, out in the rain because of my choice, rugged up or not because of my choice, eats what is given him because I choose to provide it and works when and how I choose, does not have that independence.

The Gift of Engagement

Morality and 'engagement'

Today, I drive to the yard with a pleasant sense of anticipation; one of looking forward to working and playing together. I am struck by the similarity between this feeling of personal commitment and that key equestrian concept of *engagement*. We want our horses to be engaged in the sense of physically committing themselves to their movements. We ask ourselves: 'Is he slopping into the halt, or truly engaged into it?' 'Is he lurching forward into a canter transition, or engaging behind and lifting himself into it?' 'Are his hocks underneath him as he prepares to jump?' 'Is he making that perfect V-shape as he trots – or is he, as de Kunffy so memorably puts it, "a front horse and a back horse"?' And of course we all know, deep down, that a horse cannot be truly engaged physically unless he is also engaged mentally. Which brings us closer to that other sense in which we use the word: a mutual promise of commitment between two beings who care about each other and intend to stay that way. My engagement to him makes it more likely that he will engage with me in return.

Engagement is a social, but also a moral, issue. Given a choice, it loads us one way rather than another. Ultimately, it asks us whether we are engaged with ourselves, with others, with living our lives.

I remember having a conversation years ago with a fellow-judge for whom I was writing. Our dressage class had many last-minute drop-outs and frequent periods when we had nothing to do but talk to each other. I forget what prompted our choice of topic, but we ended up talking about having to make the decision to put horses down. 'We have taken over the responsibility for their lives, so we must also take responsibility for their deaths', she said. It's a responsibility that, as a family, we have since had to take a number of times. My colleague framed the decision in such a way that I saw it not as a letting go or a wrenching apart but as part of my duty of care, and so I was able to experience that dreadfully sad decision as essentially positive rather negative.

However, engagement is not necessarily – or even often – a matter of heavy seriousness. It can make us feel joyful, playful and even bubbly. There have been times when going to the yard to ride made me feel as eager and full of excited anticipation as if I was going on a date!

The Gift of Engagement

Feeling all excited about getting ready to ride. I love the way this photo has caught the vibrant energy of looking forward, showing the balancing of my body against the weight of the tack. Riding involves balance in many ways!

My friend Marisian says she spends more time thinking about riding than she does actually doing it. On reflection, I think I do, too – and probably always have done. When I was a teenager having weekly lessons at a riding school on the outskirts of London, and later on, when I was a student, riding farm ponies in the Somerset fields and lanes, I used to come home and give my parents a moment-by-moment account of each ride: what we had done, where we had been. Sometimes I would trace our exact route for them on the map. A pleasure shared is a pleasure doubled – more than doubled, for recounting it (I am a person of words) anchored the experience in my mind. In reliving the experience I could reflect on what caused its snags or disappointments, emphasize what had gone well, and recapture those moments of the sublime which are

The Gift of Engagement

actually more frequent the more you notice them. In hearing me relive the ride, my parents could know that the pocket-money they so precisely matched each week to the cost of my riding gave me fun, stretched me in ways that only I could choose to impose on myself, and so made all our lives richer.

I am 19 again. The time Silver Queen keeps looking round and I can't work out why – until, several miles from home, I suddenly catch a glimpse of the little mother terrier, Millie, trotting close at our heels. Wondering how much stamina Millie has (she is very short-legged), I reluctantly turn back instead of completing the circuit I have intended, touched by her faithfulness and Silver Queen's careful monitoring of her presence… The time I decide to ride Rosie bareback because I am so fed up with having to get off and readjust her saddle (I am too big for her and it keeps slipping forward): the soft, squishy warmth of her against my knees. Her endless curiosity about what's behind the hedge, through the gate, what she can smell underfoot. I decide she is a village gossip. We gossip together… The rhythm of riding and the rhythm of the seasons – the familiar yet seasonally varied route along the ridge with its westward views over Somerset, into the woods where, in their time, the bluebells and primroses thicken by the path, down the long slope and up; the canter along the next ridge until the bridleway opens up onto a lane and to yet wider views. The names of the places like an incantation: Inwood, the Look-out, Broadsalls, Toomer, Milborne Wick, Bowden… (My eighteenth-century map spells them Tummer and Week.) I am talking to the old man in the village who remembers as a boy helping his uncle drive herds of over a hundred horses from the Blackmore Vale to Bristol to be sold…

Through my years of riding I am singing the First World War marching songs my father (who was 11 when that war began) used to sing me to sleep with, singing them to Rosie, Lolly, Vals, Mouse. 'It's a long way to Tipperary…', 'Keep the home fires burning…', 'Pack up your troubles in your old kit bag…'

The horses connect me to times and places beyond my own…

The primary function of morality is actually to connect you to things beyond yourself. Moral choices and moral actions are ones that have been informed by your awareness of how others could feel and experience their impact, or by a sense of how you fit into a bigger picture and

The Gift of Engagement

how this moment fits into a longer time-frame. Morality gets you expanding your perspective from your natural, inbuilt, instinctive first-person take on the world to include second-person and third-person dimensions, invites you beyond your here-and-now sensations and impulses and gets you considering causes and consequences. Morality enlarges you.

It's not that you should automatically put others before yourself; it's not that you should always be guided by potential consequences or take time to think about 'why'. It's that, if you do have these bigger issues in mind, the decisions you make and the actions you take will be richer and more grounded – even when (perhaps especially when) you put yourself rather than others first, or opt for the immediate rather than the long-term outcome. It's the sneaking sense that you are being selfish, not the actual choice you make, that undermines putting yourself first.

I talked some time ago with a rider whose ability to succeed in competition was being compromised by his very determination to perform well. Unpacking the 'loading' he put on doing well allowed him to realize that doing well made him feel justified in spending as much time and money as he did on his riding. But he had never openly checked with his family whether they resented what his hobby had been costing

Morality enlarges us. For me, this picture shows what's best about our coaching approach to helping each other: there's a real three-way conversation going on, as I believe there should be, and it makes me smile to see Ollie taking an equal part. We are all made bigger and better by negotiating with others' experience and natures.

The Gift of Engagement

them, directly or indirectly. As it turned out, his secret fears were groundless. Discussing this openly with his family allowed him to know that they were pleased for and proud of him and in no way felt resentful or deprived. And, in turn, this removed an unnecessary and debilitating pressure from his competing. It was the secrecy, not the having, that was selfish.

I have learnt a new way of asking for rein-back. The aiding I was originally taught and used until yesterday was to:

- lighten my seat by leaning slightly forward;

- put my lower legs back behind the girth and exert slight pressure;

- resist passively with my hands.

This combination was expected to tell the horse that I wanted him to move while (nicely) blocking forward movement so that he realized he had to go back and found it easier to do so because his back felt freer. It seems to me now, on reflection, that this set of instructions has two things wrong with it. First, it tells the horse to move and then tells him *not* to move in the direction he is used to and first thinks of. NLP describes this confusing process through the analogy of telling someone *not to think of a pink elephant*. Negatively framed instructions like this tend to generate puzzlement and frustration in the recipient (which may be why so many horses evidence a confusion or even crossness at being asked for rein-back). There's no inherent logic in the instructions – though the horse may well learn to obey them through learning a Pavlovian stimulus-response kind of association. Second, no specific instruction is given to the horse as to which leg he should move first. Given that two-legs-on signifies 'move forward' it's hardly surprising if, on initiating forward movement and finding it blocked by resisting hands (however passive rather than active they may be) the horse begins to move one leg at a time – which in dressage rein-back constitutes a fault. As so often, we get what we ask for.

As a judge, I have watched and marked many, many hundreds of rein-backs. The very word tells you that the reins are an important part of the procedure, even though everyone 'knows' you mustn't pull. In competition you can see horses inching back, single foot by single foot rather than by diagonal pairs; you can see some hollowing and resisting

The Gift of Engagement

the pull/signal of the reins; you can see others ducking and running back behind the expected pull of the reins; you can see them lurching sideways into crookedness. Only rarely do you see them calm, free, obedient, straight, offering clearly diagonal steps, and accepting the bit just as easily as they might do if they were going forward. Good rein-backs don't come any more easily at higher levels than lower – which I now realize may well be the result of our imprecise aiding. To make matters worse, rein-back is sometimes even used as a reminder of subservience, as both dressage riders and showjumpers often halt then rein-back before they start their competition round.

The hidden moral issue with rein-back is: *for whose benefit are we reining-back at all?* In the stable, your horse needs to be polite enough to back away from his door to allow someone to come in to groom, tack-up or muck-out. Out hacking, he needs to rein-back at times to help you open and shut gates. These situations both carry the inherent logic of practical need. Otherwise, rein-back is advocated as one of many suppling and gymnasticizing exercises, yet only rarely does it help to achieve this aim.

Yesterday, my friend Debbie showed me how she aided her Advanced horse, using a pattern of signalling she had evolved when she first had him because he was so sharp, strained and afraid that he would only accept refined and subtle aiding (wise horse!) She took account of him, listened to him, made it simple and easy for him – made it step-by-step. First, she establishes a forward-thinking, engaged, committed halt. No change then in the seat, no change in the weight of the reins. Both legs go back to alert him, then alternate squeezes from Debbie's legs cue the movement of his alternate pairs of legs. One squeeze per step makes it easy to inform him exactly how many steps are wanted, and makes it easy for him to oblige without anxiety, hesitation, resistance or rushing. There are no mixed messages, only clarity and understanding.

Today, I am riding Mouse. He has had a day and a morning off in the field, and is not very attentive. I am tired because I've ridden three days running. Our harmony is not at its peak. I explain to Nikki, who is in the school at the same time lungeing her horse, what I learnt about rein-back yesterday. Not really expecting much, I halt Mouse and experiment. It seems fair to lighten my seat, since this is what I think he's used to, but I'll see what happens if I give alternate rather than simultaneous

The Gift of Engagement

leg-aids. He hesitates, clearly puzzled. But then I sense 'the try' as he shifts his weight just a little. Pat, pat. 'Good Mouse'. We move forward and then I try again. We get a couple of steps. More pats. Walk on, find another place, make another halt. I am concentrating on trying to discover just how much pressure I need, so I forget to lighten my seat. Bingo! By mistake, my message becomes so much simpler for him. Backwards – of course we can! 'How many steps do you want, Mum?' He is clearly so pleased to have understood what I wanted, and so pleased to be able to do it. Thoughtful, empathic Debbie. Clever Mouse.

Through their astonishing sensitivity, horses show us that thoughts are things. Therefore, just as a correct physical position which is in harmony with the laws of physics is indispensable to allow essential energy patterns to flow through our own and the horse's body; so too, a sound mental position – constructive, wholesome, compassionate thinking which is aligned with the benign energy of cosmic law – enables it to flow through and strengthen all our endeavours… I believe it is worthwhile to review briefly some of the most common personal obstacles to success in horsemanship: immodest ambitions, arrogance, anger, irritability, having dictatorial or judgmental attitudes, impatience, lacking self-control (especially over our emotions), fear, anxiety, being inconsiderate, bearing grudges, resentment, or kindling vengeance in our hearts.

Erik Herbermann, *A Horseman's Notes*, p.13

Two horses passing each other, both in his own way engaged with his rider and with his work. And as if that weren't enough, each horse is being ridden by the other's owner… Though invisible here, we are engaged, too.

The Gift of Engagement

In writing this book, I find myself realizing that each chapter has its own underlying theme, and that the themes are actually qualities of living. It seems to me that, together, they make up the essential qualities of a life well-lived.

Opportunity	Poise
Reflection	Self-acceptance
Presence	Transcendence
Spaciousness	Engagement
Influence	

These are the gifts our horses offer us every day – if we will only accept them.

The Gift of Engagement

Coda: Four Days on the Trot

I realize that there is a little more yet to say, and another gift to celebrate and be thankful for. Engagement is the promise to commit: connection is the fruit it brings us.

Connection is both a physical reality and a spiritual experience. Physically, it's what you get when your horse willingly offers you the energy of his whole body without blocking or withholding – and when you guide and receive it through the consistent and selfless ministrations of your seat, legs and arms. Mentally, it's that consistency and subtlety of mutual attention that confirms moment by moment 'I'm with you', each to each. Emotionally, spiritually, it's the wordless attuning that means you have only to *think* for your thought, your feeling, to be understood – either way. Connection between horse and rider, as in any good relationship, is about empowering each other to be the best you can.

Of course it's rare – but not that rare. It's the promised, sought-after, wondered-at, relished, remembered peak of our shared experience. It's what makes us smile in every cell. It's the goal that keeps us going because we 'know in our bones' that it's achievable because it's about how we *are*, not just what we can *do*.

Here's a vignette of four recent consecutive days that showed me what I mean. I hope that, as you imagine yourself into my experience, it will reflect back to you something special of your own.

Day 1

Nikki has volunteered to give me a hand with Mouse – by which we mean that it's her turn to take the coaching role in one of our 'swop sessions'. Mouse has just spent a day and a night in the field and, to start with, plods round like an old donkey. This is not how he was last week – nor how Nikki reported him as being when, in my absence, she rode him at the weekend. Then, he was apparently offering to be round and 'through' from the beginning... Today, his mind and his body are not on offer. If I want them, I shall have to work for them.

I feel 'all arse and elbows' as I try to ask him to go forward. Nikki and I have decided that this is the only way to get him engaging instead of diddling along with his hind legs, and the only way to get a jump and a true three-beat in canter instead of the four-beat he offers when he lacks animation and energy. Nothing in Mouse seems to be consistent today – and nothing in me – either. My legs are everywhere as I press (and kick) him forward, my feet slipping home in the stirrups like a beginner's. When I bend my elbows and frame a contact for him to work into, he sticks his head in the air; when I offer the reins forward he goes flat, or briefly dips his head only to lift it again as soon as I try to resume a tentative connection with his mouth. It's tiring and dispiriting.

For the moment we abandon the prospect of roundness and 'throughness' and work simply on getting him going forward. He feels sticky with reluctance – a disengagement of his mind which is making him jib and stutter, cutting in so that he is hard to steer, threatening to nap or even stop entirely. 'Stay with it', Nikki encourages me. She suggests changing back and forth between trot and canter. At least we are getting more jump in the canter. After a while there's some spark in the trot. After about half an hour, Mouse finally offers to lower his head and raise his back – grudgingly, but enough for us to feel we haven't all entirely wasted our time.

Day 2

I have been wondering what to do today after yesterday's disappointment. I went home feeling incompetent and deeply inelegant. I know I can create a secure contact and ride a horse into it – but there seemed to be no starting-place for that yesterday, no way to achieve it. Last night I said to Leo: 'I hope Mouse feels his muscles as much as I do!' I felt a long way from my ideals of empathic, clear communication, mutual understanding and pleasure in partnership. Today, I am physically tired and emotionally flat. Should I try to do more of the same? Or would that be boring and frustrating again for us both? I don't want another battle. If I opt for work in walk, or lateral movements, or a slow balanced trot, would that help restore our harmony – or would it be a cop-out for me and no advance for him?

As I am tacking-up, Nikki arrives: 'I thought I might have a go on your horse this afternoon, if that's all right with you', she says. Gratefully, I make a coffee to take down to the school, but I do pick up my helmet so that I can ride a bit…

I feel relieved and strangely exonerated when Mouse starts out with no more apparent willingness for Nikki today than he offered yesterday to me. He is doing his characteristic 'camel impression'. But at least he does seem to be going forward with more energy, even though Nikki says she is having to work hard for it. She repeats yesterday's pattern of alternating trot and canter several times, with a loose rein so he doesn't have anything to fight against. And there is no napping. After about fifteen minutes she 'makes her arms and hands into a pair of side-reins': there is weight and passive presence to them, but no pulling back and no fiddling. At last he seems to understand that the energy he is producing with his hind legs can bow up through his back and allow him to round through onto the bit. Then there is a connection… Then there is a trot and a half… Nikki is smiling and telling Mouse, 'Good lad…good lad'. I am grinning and calling out 'That's fabulous! … Now those hind legs are working! … Now you've even got some suspension!'

The Gift of Connection

A trot and a half. This is the Mouse that can be...

They go forward into canter: it's big, round, true, hugely bounding and ground-covering. 'I could sit on this all day', she calls out. More trot again, just as connected and powerful as before. Then: 'Now you sit on him and feel it.' I make side-reins of my hands and arms. I encourage (not drive, today) with my legs. We have that trot again. 'He's giving you even more than he gave me', Nikki says. I need and feel the confirmation, both from her and from Mouse. It's fabulous. He's fabulous. And of course I can ride well after all – I'm fabulous too! Still hands, straight back, close, supporting legs, powerful forward rising to open and support his big balanced steps...still hands, forward hands, framing, giving, thinking hands. I can't get over my hands. It's what I have been dreaming of. It's what I knew – believed – hoped – could happen.

The Gift of Connection

Day 3

I have asked Gaz if he will come for a hack with us today. We deserve something less pressured. My fearful self suggests that we have a gentle ride (by which I mean maybe a road ride, maybe a ride at walk…). I am kidding myself that we deserve this after two energetic sessions. Gaz, who always says he is not a psychologist (!), tells me I'm not getting away with that! We need more – for Mouse's sake. Huh! We ride through the fields. I decide that Mouse is not going to do his riding-school horse impression and tag along behind Gordon. Today he is going to walk alongside his mate – and he is going to stride out enough to keep pace with Gordon's quicker steps without having to put in bursts of trot to catch up as he usually does. It feels companionable. It feels more purposeful. It feels good.

Gaz inveigles me into a very polite collected canter over the stubble. Mouse is balanced, rocking-horse, on the bit. We walk along the next side of the field, then turn back across it again. Another canter. As we crest the rise Mouse lengthens his stride, which instinctively alarms me because I didn't ask for it. 'It's only because you have started to go downhill', Gaz says. 'He will always come back to you' – and he already has. Then we turn again and trot where we first cantered – just to make the point (not to Mouse, I know, but to me) that I am the one who decides – and he is the one who is happy to agree.

In the driveway leading back towards the farm Gordon spooks at washing which has been hung out over the lodge gates to dry. When I drove past on my way to the yard I saw the large swathes of fabric hanging there and anticipated that Mouse might object to them if we went that way – but now when I put my legs on firmly he strides calmly past, giving his more flighty mate Gordon a lead.

Day 4

We arrive at the yard at the same time as Nikki, so she and I can encourage each other by riding in the school at the same time. Today, Leo has brought his camera. He has been unwell for a month with a debilitating inner-ear infection, and this is the first time he has left the house. He says that he feels 'convalescent', but this doesn't seem to stop him scurrying back and forth around the sand school taking photos. Being around the horses again after all this time seems to be restoring and energizing him.

I am keen to replicate the work that Nikki and I did the day before yesterday with Mouse – but today it will be down to me. It really does seem that we have found a way of pacing and then leading Mouse, so that he can begin by feeling free to go forward and is only asked to pick himself up and put himself together when he feels truly ready. As with pacing and leading in my human coaching, I have to sensitize myself to notice those moments when he makes a small 'try', and to praise and capitalize on this without grabbing at it. And I need both the patience and the consistency to encourage Mouse to convert his initial 'try' into a more fully developed offering.

We get the 'try', and the offer. We get sustained connection and harmony. I'm excited and triumphant – and about to become so in a second way. At intervals today I have been noticing the lovely work that Nikki and Ollie are doing together, and shouting appreciation to them. It's so good to see Ollie now: when Nikki first had him he was tense, drawing back from the contact, anxious and sometimes quite spooky. Now, after all Nikki's patient encouragement, he is not only established in his work but happy in himself and his rider. Nikki calls out 'Why don't you sit on him – let's swop.'

I haven't sat on Ollie for a couple of years. Immediately I feel there's a major difference in him. There's also a major difference in me. Ollie is confirming with every moment that I can indeed ride. His dressage is more advanced than Mouse's, as is his balance – but what excites me is not the technical stuff but the responses that tell me he understands what I want and is willing to oblige. He is light in the hand – I have that 'reins of silk thread' feeling; he is so on my seat that I have only to think 'collection' and he gives me the most elegant downward transitions. I even have the courage to ask him for canter (my old dread on unfamiliar horses!), and he is light and 'through' and attentive for me. What a joy it's been to partner these two splendid and generous horses.

While you can tell from my forward bend and rather 'gripping up' legs that I've not entirely lost my anxiety about cantering unfamiliar horses, you can also tell from our faces and from Ollie's quiet mouth that we are okay working together. There's a sense of peaceful accommodation between us that I like: it really did feel like that at the time.

The Gift of Connection

Riding Ollie has somehow provided me with the evidence and the culmination of all that I have been writing about, because he and I have no agenda and no history with each other. For me to be so unusually calm and bold and him to be so responsive, there have to have been deep changes in how I think about myself and how I feel in myself. Today's riding has brought everything together for me and in me. What I have learnt from my frustration, my experimentation, and my struggles with myself and Mouse has become part of who I am. It is 'in my muscle' now. Ollie has been the proof of that. As Belasik says in *Riding Towards the Light* (pp.126–7): 'I would like to be able to say that immediately after my experience… I was overcome with joy and celebration… The true feeling was more like a sense of peace… I finally perceived that everything was really just beginning.'

What's it all about? It's about the fact that every day is different. That we have to ride every day according to what our horse, our own state and the external world present us with. It's about the fact that horses are sentient and changeable too – we can't assume they will be the same

Two plus two is more than four. What a good time we each had, and how much we have enjoyed each other's company.

The Gift of Connection

today as yesterday, in either a good or a bad sense. It's about the fact that the authenticity and solidity they possess in themselves can become available to us if we interact with them openly, honestly and with love. It's also about the fact that although they are our friends they are *not* actually people – they have horsy not human reactions and agendas, which we must take into account in their terms. It's about the hard fact that the bad days are as much a gift as the good ones – if we have the grace to accept them and the wit to profit from them. It's about the fact that it takes two. It's about the quality of the connections.

A winter's day, and a beautiful friendship. This picture of Roy and Cori in the rain, with its soft focus and rain-spot brilliance, implies so much that is special, and yet hard to define in words.

Bibliography

Argyris, Chris and Schon, Donald, Theory in Practice: *Increasing Professional Effectiveness*, Jossey-Bass (San Francisco) 1974.

Belasik, Paul, *Riding Towards the Light*, J. A. Allen (London) 1990.
—— *Dressage for the 21st Century*, J.A. Allen (London) 2002.

Budiansky, Stephen, *The Nature of Horses*, Weidenfeld & Nicholson (London) 1997.

Cleese, John and Skynner, Robin, *Families and How to Survive Them*, Methuen (London) 1983.

Deering, Anne, Dilts, Robert and Russell, Julian, *Alpha Leadership* (Wiley, New York) 2002.

De Kunffy, Charles, *The Ethics and Passions of Dressage,* Half Halt Press (Middletown, Maryland) 1993.

Dilts, Robert, *Changing Belief Systems with NLP*, Meta Publications (Capitola, California) 1990
—— *Visionary Leadership Skills: Creating a World to Which People Want to Belong*, Meta Publications (Capitola, California) 1996.
—— *From Coach to Awakener*, Meta Publications (Capitola, California) 2003.

Gallwey, Timothy, *The Inner Game of Tennis*, Pan (London) 1974.
—— *The Inner Game of Golf*, Pan (London) 1986.

Grandin, Temple and Johnson, Catherine, *Animals in Translation*, Bloomsbury (London) 2005.

Herbermann, Erik, *Dressage Formula*, J.A. Allen (London) 1980 (new edn 2008).
—— *A Horseman's Notes*, Core Publishing (Crofton, Maryland) 2003.

Kiley-Worthington, Marthe, *Horse Watch, What it is to be Equine*, J.A. Allen (London) 2005.

Kohanov, Linda, *Riding between the Worlds*, New World Library (Novato, California) 2003.

Maslow, Abraham, *The Farther Reaches of Human Nature*, Pelican (Harmondsworth, Middlesex) 1971.

McCormick, A.V.R. & McCormick M.D., *Horse Sense and the Human Heart*, Health Communications, Inc. (Deerfield Beach, Florida) 1997.

Naish, John, *Enough: Breaking Free from the World of More*, Hodder & Stoughton (London) 2008.

O'Connor, Joseph & McDermott, Ian, *The Art of Systems Thinking*, Thorsons (London) 1997.

Postman, Neil & Weingartner, Charles, *Teaching as a Subversive Activity*, Penguin (Harmondsworth, Middlesex) 1969.

Rashid, Mark, *Horses Never Lie*, Johnson Books (Boulder, Colorado) 2000.

—— *Life Lessons from a Ranch Horse*, David & Charles (Newton Abbot) 2004.

—— *Horsemanship Through Life*, Johnson Books (Boulder, Colorado) 2005.

Roberts, Monty, *The Man who Listens to Horses*, Arrow Books (London) 1997.
—— *Join-Up, Horse Sense for People*, Harper-Collins (London) 2000.

Rogers, Carl, *On Becoming a Person*, Constable (London) 1967.

Rossi, Ernest, *The Psychobiology of Mind-Body Healing*, Norton (New York and London) 1986.

Strozzi, Ariana, *Horse Sense for the Leader Within*, Author-House (Bloomington, Indiana) 2004.

Truax, Charles B. and Carkhuff, Robert R., *Toward Effective Counselling and Psychotherapy: Training and Practice*, Aldine Atherton (Chicago & New York) 1967.

Webb, Wyatt with Pearlman, Cindy, *It's not about the Horse*, Hay House, Inc. (Carlsbad, California) 2002

Winnicott, D.W., *Playing and Reality*, Tavistock (London) 1971.

Permissions

The author and publisher are grateful to the following for their kind permission
to reproduce material: Ariana Strozzi for permission to quote from *Horse Sense for the
Leader Within*, Health Communications Inc. for permission to quote from *Horse Sense
and the Human Heart* by McCormick & McCormick, Constable for permission to
quote from *On Becoming a Person* by Carl Rogers, Methuen Publishing for permission
to quote from *Families and How to Survive Them* by John Cleese and Robin Skynner,
Meta Publications for permission to quote from *Changing Belief Systems with NLP,
Visionary Leadership Skills: Creating a World to Which People Want to Belong* and *From
Coach to Awakener* all by Robert Dilts, Pan Macmillan for permission to quote
from *The Inner Game of Tennis* by Timothy Gallwey (© Timothy Gallwey, 1974),
WW Norton & Company, Inc. for permission to quote from *The Psychobiology of
Mind-Healing* by Ernest Rossi, Johnson Books (Boulder, Colorado USA) for
permission to quote from *Horses Never Lie* and *Horsemanship Through Life* both by
Mark Rashid, New World Library (Novato CA) for permission to quote from *Riding
Between the Worlds* (© 2003) by Linda Kohanov, Hay House, Inc. for permission to
quote from *It's Not about the Horse* (© 2002) by Wyatt Webb, The Orion Publishing
Group Ltd. for permission to quote from *The Nature of Horses* by Stephen Budiansky,
Penguin Group (USA) Inc. (New York) for permission to quote from *The Farther
Reaches of Human Nature* by Abraham Maslow and Bloomsbury Publishing for
permission to quote from *Animals in Translation* by Temple Grandin.

Despite our best efforts, we have been unable to ascertain definitive ownership of the
copyright of some material in this book. If anyone feels they have a claim, please
contact the publisher.

alance and beauty challenge and companionship completeness
nd exercise fun and fulfilment fear and frustration motivation a
ecovery radiance and reassurance softness and sensitivity
hallenge and companionship completeness and connection dr
nd fulfilment fear and frustration motivation and movement plec
nd reassurance softness and sensitivity satisfaction and sc
ompanionship completeness and connection drive and desire
ear and frustration motivation and movement pleasure and
eassurance softness and sensitivity satisfaction and solace war
ompleteness and connection drive and desire energy and enthu
otivation and movement pleasure and pain peace and purpos
ensitivity satisfaction and solace warmth and wonder bala
onnection drive and desire energy and enthusiasm excitement
ovement pleasure and pain peace and purpose relaxation
atisfaction and solace warmth and wonder balance and beauty
nd desire energy and enthusiasm excitement and exercise fun
nd pain peace and purpose relaxation and recovery radian
armth and wonder balance and beauty challenge and compar
nthusiasm excitement and exercise fun and fulfilment fear and
urpose relaxation and recovery radiance and reassurance s
alance and beauty challenge and companionship completeness
nd exercise fun and fulfilment fear and frustration motivation a
ecovery radiance and reassurance softness and sensitivity satis